Rethinking Order

Also available from Bloomsbury

God and the Meanings of Life, T. J. Mawson
Nothingness and the Meaning of Life, Nicholas Waghorn

Rethinking Order

After the Laws of Nature

Nancy Cartwright and Keith Ward

BLOOMSBURY ACADEMIC
LONDON • NEW YORK • OXFORD • NEW DELHI • SYDNEY

BLOOMSBURY ACADEMIC
Bloomsbury Publishing Plc
50 Bedford Square, London, WC1B 3DP, UK
1385 Broadway, New York, NY 10018, USA

BLOOMSBURY, BLOOMSBURY ACADEMIC and the Diana logo are
trademarks of Bloomsbury Publishing Plc

First published in Great Britain 2016
Paperback edition published 2019

A catalogue record for this book is available from the British Library.

ISBN: HB: 9781474244060
PB: 9781350089891
ePDF: 9781474244053
ePub: 9781474244084

Names: Cartwright, Nancy, editor.
Title: Rethinking order : after the laws of nature / edited by Nancy
Cartwright and Keith Ward.
Description: New York : Bloomsbury Publishing Plc, 2016.
Identifiers: LCCN 2015034813| ISBN 9781474244060 (hb) | ISBN 9781474244053
(epdf) | ISBN 9781474244084 (epub)
Subjects: LCSH: Order (Philosophy) | Philosophy of nature. | Cosmology.
Classification: LCC B105.O7 R46 2016 | DDC 117–dc23 LC record available at
http://lccn.loc.gov/2015034813

Typeset by Fakenham Prepress Solutions, Fakenham, Norfolk NR21 8NN

To find out more about our authors and books visit
www.bloomsbury.com and sign up for our newsletters.

Contents

Acknowledgements

After the John Templeton Foundation, which gave us funds to study the big question of concepts of order that do not rely on laws of nature, then left us to carry on the research as the emerging ideas suggested, the participants in the Order Project would most like to thank the project administrator, Rebecca Robinson, who ensured that facilities, time and arrangements for conducting the research and disseminating it were all in place where and when needed.

Contributors

William Bechtel is Professor of Philosophy and a member of the Center for Chronobiology at the University of California, San Diego. His research explores issues in the philosophy of the life sciences, including systems biology, cell and molecular biology, biochemistry, neuroscience and cognitive science. He has published several books, including *Discovering Cell Mechanisms: The Creation of Modern Cell Biology* (Cambridge 2006) and *Mental Mechanisms: Philosophical Perspectives on Cognitive Neuroscience* (Routledge 2008), as well as numerous articles in journals such as *Philosophy of Science, Biology and Philosophy* and *Studies in History and Philosophy of Biological and Biomedical Sciences*.

Robert C. Bishop is Professor of Physics and Philosophy as well as the John and Madeleine McIntyre Endowed Professor of Philosophy and History of Science at Wheaton College. His research involves history and philosophy of science, philosophy of physics, philosophy of social science, philosophy of mind and psychology, and metaphysics. He has authored a book, The *Philosophy of the Social Sciences* (Continuum 2007), as well as several articles published in *Analysis, Philosophy of Science, Foundations of Physics* and *Studies in History and Philosophy of Modern Physics*, among other journals.

John Hedley Brooke taught the history of science at Lancaster University from 1969 to 1999. In 1995, with Geoffrey Cantor, he gave the Gifford Lectures at Glasgow University. From 1999 to 2006 he was the first Andreas Idreos Professor of Science and Religion at the University of Oxford, Director of the Ian Ramsey Centre and Fellow of Harris Manchester College. After retirement, he was designated

'Distinguished Fellow' at the Institute of Advanced Study, University of Durham (2007). He has been the President of the British Society for the History of Science, of the Historical Section of the British Science Association, of the International Society for Science and Religion and of the UK Forum for Science and Religion. Among his books are *Science and Religion: Some Historical Perspectives* (1991, 2014), *Thinking about Matter* (1995) and (with Geoffrey Cantor) *Reconstructing Nature: The Engagement of Science and Religion* (1998). His most recent book, co-edited with Ronald Numbers, is *Science and Religion Around the World* (2011).

Nancy Cartwright FBA is Professor of Philosophy at the University of Durham and at the University of California, San Diego. She is former president of the Philosophy of Science Association and the American Philosophical Association (Pacific) and a member of the German Academy of Sciences (Leopoldina) as well as a MacArthur fellow and Fellow of the British Academy and of the American Philosophical Society (the oldest US academic honorary society). Her research interests include philosophy and history of science (especially physics and economics), causal inference, objectivity and evidence-based policy. She has authored several books, including *The Dappled World* (Cambridge 1999) and *Evidence-Based Policy: A Practical Guide to Doing it Better*, with Jeremy Hardie (Oxford 2012), as well as numerous articles published in venues such as the *Journal of Philosophy*, *Philosophy of Science* and *The Lancet*.

John Dupré is Professor of Philosophy of Science at the University of Exeter, and Director of Egenis, the Centre for the Study of Life Sciences. He has formerly taught at Oxford, Stanford and Birkbeck College, London. His publications include: *The Disorder of Things: Metaphysical Foundations of the Disunity of Science* (1993); *Human Nature and the Limits of Science* (2001); *Humans and Other Animals*

(2002); *Darwin's Legacy: What Evolution Means Today* (2003); *Genomes and What to Make of Them* (with Barry Barnes, 2008); and *Processes of Life: Essays on the Philosophy of Biology* (2012). He is a former President of the British Society for the Philosophy of Science and a Fellow of the American Association for the Advancement of Science.

Roman Frigg is Director of the Centre for Philosophy of Natural and Social Science, Associate Professor of Philosophy and Co-Director of the Centre for the Analysis of Time Series at the London School of Economics and Political Science. His current work focuses on predictability and climate change, the foundations of statistical mechanics and the nature of scientific models and theories. He has published several articles in journals such as *Philosophy of Science*, *Erkenntnis* and *Synthese*.

Rom Harré is Distinguished Professor of Psychology at Georgetown University and former director of the Centre for Philosophy of Natural and Social Science at the London School of Economics and Political Science. His current research is in social psychology though he worked for a long time on topics in metaphysics and the philosophy of science. He has published a large number of books, including *Pavlov's Dogs and Schrödinger's Cat* (Oxford 2009), as well as numerous articles in both philosophy and psychology journals.

Steven Horst is Professor and Chair of Philosophy at Wesleyan University. His research is in philosophy of psychology, moral psychology and metaphysics. He has published two books, including *Beyond Reduction: Philosophy of Mind and Post-Reductionist Philosophy of Science* (Oxford 2007), and several articles that have appeared in *Synthese*, *Minds and Machines* and *Philosophical Psychology*, among other journals

Eric Martin is Assistant Professor of History and Philosophy of Science in the Honors College at Baylor University. His research is in general history and philosophy of science, philosophy of biology and the relationship between religion and science. He has, among other articles, authored an article on 'Evil and Natural Science' with Eric Watkins, for the collection *Oxford Philosophical Concepts: Evil – A History*, edited by Andrew Chignell (Oxford forthcoming).

T. J. Mawson is a Fellow and Tutor in Philosophy at St Peter's College, Oxford University. His research interests are the philosophy of religion, philosophical theology and moral philosophy. He has published two books, *Belief in God: An Introduction to the Philosophy of Religion* (Oxford 2005) and *Free Will: A Guide for the Perplexed* (Continuum 2011), and a number of articles in journals such as *Religious Studies*, the *British Journal for the History of Philosophy* and *Think*.

Eleonora Montuschi is Associate Professor in the Department of Philosophy and Cultural Heritage at the University of Venice and Senior Research Fellow at the London School of Economics and Political Science. She is a philosopher of the natural and social sciences working on scientific objectivity, the theory and practice of evidence and methodological issues of the social sciences. She has published a book, *The Objects of Social Science* (Continuum 2003), as well as various articles in journals such as *Axiomathes*, *Theoria* and *Social Epistemology*.

Russell ReManning is Reader in Philosophy and Ethics at Bath Spa University. His research is in theology, the philosophy of religion, the relationship between religion and science and the thought of Paul Tillich. He has edited several books, including *The Oxford Handbook of Natural Theology* (Oxford 2013) and *The Cambridge Companion to Paul Tillich* (Cambridge 2009).

Keith Ward FBA is a Professorial Fellow at Heythrop College after having held positions at Cambridge University, Gresham College (where he was the Gresham Professor of Divinity), Oxford University (where he was Regius Professor of Divinity and Canon of Christ Church) and the University of London (where he was Professor of the History and Philosophy of Religion), among others. He was ordained Priest of the Church of England in 1972. His research is in theology, history of philosophy, philosophy of religion, metaphysics and the history of religions. He has published numerous books, including *Kant's View of Ethics* (Blackwell 1972), *Rational Theology and the Creativity of God* (Blackwell 1984) and *Comparative Theology* in five volumes (Oxford 1994–2000 and SCM Press 2008).

Eric Watkins is Professor of Philosophy at the University of California, San Diego. His primary area of research is Kant's philosophy. He also works on early modern philosophy (Leibniz, Newton, Hume), German idealism (Reinhold, Fichte, Schelling and Hegel) as well as on the history of philosophy of science. He has authored *Kant and the Metaphysics of Causality* (Cambridge 2005) and edited *The Divine Order, the Human Order, and the Order of Nature: Historical Perspectives* (Oxford 2013), among others. He has also published numerous papers in journals such as the *Journal of the History of Philosophy*, *Studies in History and Philosophy of Science* and *Philosophy and Phenomenological Research*.

Introduction

Keith Ward

This book has one central argument. The argument is that the classical Newtonian worldview of a universe made of material particles governed by absolute and unbreakable laws of nature is obsolete. Of course, this has been known since the first formulations of quantum mechanics in 1925, but its implications have rarely been taken to heart by the general educated public. Also, this revolution in our understanding of the universe is not just a matter of quantum physics. Even more basically, it is a matter of our understanding of the basic laws of nature, and of how nature comes to have the sorts of order, integration and organization that it does have. That understanding has changed, and this book tries to show how it has changed. To put it in a brutally simple way, there are no absolute and unbreakable laws of nature whatsoever which determine exactly how all things will happen, so that, as the French physicist LaPlace put it, if we knew the initial physical state of the universe, and all the laws of nature, we would be able to predict everything that would ever happen.

Albert Einstein was never able to accept this. He is reported to have said 'God does not play dice with the universe', and he believed with fervent faith that there must be deterministic laws underlying our observed universe, even if we humans would never be able to use them to make absolutely correct predictions. I think we should applaud Einstein's faith – it is just the sort of thing that makes truly great scientists – but we have to say that it was probably mistaken. We can even say that there was never a very good reason to think it was correct. Why should there be unbreakable deterministic laws, after all? We, the writers of this book, do not think there are no laws of nature at all. The physical world is highly ordered, and manifests organized complexity at many levels. It is not all chaos and accident. But the principles of order in nature are much more local, diverse, piecemeal, emergent and holistic than the old model of one absolute set of laws (which would be, basically, the laws of the 'master science', physics) dictating the behaviour of fundamental particles in such a way that, in principle, the behaviour of all complex biological, personal and social entities can in principle be deduced from them.

This view can be expressed in two main ways. We can say that there are laws of nature, but those laws are not what we thought – not absolute, unbreakable and forming a closed causal system that can completely explain everything about the world. Or we can say that 'law' is not the best model for explaining the order that exists in nature. Perhaps a different model like that of 'power', 'capacity' or 'disposition' might be a better one. Objects might, for instance, have various capacities which may be realized or frustrated in different local contexts, and which are sensitive to novel background conditions which have never existed before.

Nancy Cartwright has been influential in urging such a view, which she often presents by speaking of 'science without laws'. She certainly does not mean that nature is without order – though that order may be less all-encompassing than we think – but the term 'law' (which is, after all, a metaphor) might not be the best term for describing that order, especially if laws are thought to be unbreakable, universal and completely explanatory of all that occurs in nature. All of the writers of this book broadly agree with her, though of course they do not all believe exactly the same things. So we are arguing for a new perspective on the way nature manifests order and intelligibility.

What difference will this change of perspective make? There will be many specific differences in particular sciences, but the most obvious overall change will be the collapse of the model of science as one unified and totalizing programme for answering every possible question about the universe. It will undermine a view of physics as the one master-science to which all others reduce in the end. It will destroy the view of the physical universe as a closed causal system into which talk of human freedom and dignity fits only with great difficulty, and which God can only interact with by breaking God's own inviolable laws of nature. It will enable us to see the universe as open, entangled, emergent and holistic, rather than as a piece of clockwork which follows a set of blind impersonal laws. The universe will be open and emergent, in that it does not run down eternally pre-set rail tracks, but inherently contains the potential for new creative endeavours. It will be entangled and holistic, in that the causal inter-relations and influences between entities are many and varied, and not reducible to just one sort of quasi-mechanistic form of causality. The emergence of new systems and contexts will have causal influence on the 'lower' and simpler component parts of those systems.

People will still take various views of whether there is objective purpose in the cosmos, on whether or not there is true human freedom and responsibility and on whether or not there is a God or something like a God. But the new perspective will enable those who see signs of objective purpose and value in the universe to feel that their views resonate positively with much state-of-the-art thinking in the sciences, and it will make it easier to see the universe as 'friendly' to both mind and consciousness, rather than that they are unanticipated and accidental by-products of mechanistic processes.

These are radical changes in our view of the universe. But this book aims to show that they are not the result of either unreasoned and sentimental faith or purely abstract theory. The change of view comes from hard thinking about the actual practice of the natural sciences as well as in philosophy and theology. No one person has the competence to cover all these fields with authority. So we have sought to find experts in physics, biology, social science, philosophy and theology who can contribute to our argument in their own way. Nevertheless, we have sought to construct a book which pursues its argument with one voice and in a systematic way. To that end, between each chapter a short editorial link has been inserted which makes explicit what is being asserted at each stage, and how it builds up into one continuous and cumulative case.

The book arises out of a four-year project sponsored by the Templeton Foundation. Sad to say, there are those who think that such sponsorship is a sign of intellectual weakness. We can firmly state that at no time has there been any attempt by Templeton to make us choose writers they like or whose views they agree with, or to influence any of those views. All the writers (who have various moral and religious opinions) formed their views long before they had heard of Templeton, and we all believe that these views are exciting and important. We are therefore very grateful for Templeton Foundation sponsorship, which has enabled us to meet and develop our views by dialogue and discussion in ways that would otherwise have been very difficult.

Part One

The Historical and Philosophical Setting

1

The Rise and Fall of Laws of Nature

Eric Watkins

Editorial Link: The book begins with a chapter by Eric Watkins which
sets the historical context for the discussion of laws of nature, and sets
out some of the alternative models for understanding order in nature
which subsequent chapters will examine in more detail. KW

It is a basic fact, if not about the world itself, then at least about us,
that *order* is crucial. Whatever our specific commitments may be,
and they obviously vary considerably, we are deeply committed to
some notion of order. Whether that is a natural order, a moral order
or a divine order, we firmly hold onto the idea that reality is, must
or, at the very least, ought to be ordered in some way. The contrast
concept – disorder or, if taken to the extreme, chaos – has undeniable
negative connotations for us, as is clear from the dismay caused by the
inconvenient placement of an 'out of order' sign. This kind of negative
attitude towards the lack of order is common, not only in the mass of
papers strewn across my desk, but also in the fundamental structure
of our social lives, since chaos and disorder are destructive of our most
basic abilities and opportunities, and the corresponding demand for
'law and order' is undeniable even for the most liberal-minded person.

It is also a basic fact, though now historical in character, that our
conceptions of order have shifted in fundamental ways over time.
Indeed, part of the point of historical inquiry is to determine how
order took different shapes at different times. For example, what

nature, or the natural order, was understood to be was radically different after the discoveries of seventeenth-century scientists such as Galileo and Newton than before. Not only did they replace a geocentric with a heliocentric astronomy such that human beings were no longer firmly entrenched at the centre of the universe, but, more generally, they also rejected qualitative explanations of the world in favour of mathematically precise quantitative descriptions of matter in motion, a change in the practice of science that has been with us in one form or another ever since. Indeed, historians of science have argued that these radical shifts concerning our fundamental conception of what nature is, and how it can be described properly, were tantamount to a Scientific Revolution. It is plausible to think that another such shift occurred with Einstein's groundbreaking discovery of special and general relativity theory and with the emergence of quantum physics in the twentieth century. The consequences in these cases are a radical change in our very conception of the spatial-temporal fabric of the universe and a different understanding of what the ultimate constituents of the world are, e.g. with strings rather than things.

What is striking in both of these cases, but in others as well, is that a radical change in our overall conception of nature and of our place in the world occurs due to a change in some very specific part of it, in this case, in the details of particular scientific theories. That is, a few very detailed modifications force a broader change at a much higher level of generality, which affects our most basic understanding of the world and our place in it. Indeed it is only when these broader changes occur that we say not simply that we've discovered some new feature of the world, but that the very order of the world is different. It is thus important to appreciate that our conception of the world includes both more general and more specific levels and that, at least in some cases, the former depend on the latter in fundamental ways and not always in a manner that is immediately transparent to us.

What this constellation of basic facts suggests is that, to understand our current situation properly, we must constantly be attuned both to the very specific changes that are happening at an ever-accelerating pace in our globalized world and to the broader implications that they might have for our more general conception of the order of things. But what do I mean when I refer to 'our more general conception of the order of things'? After all, there are many different kinds of order, and this is true even when order is considered at a very general level: one can speak of a spatial order, a temporal order, a logical order or even, for that matter, a pecking order. What is common to them all is that they indicate some kind of abstract principle of organization, some way in which a plurality of things of one kind are related to each other in a way that we find to some degree intelligible or harmonious. This is by no means meant to be an exhaustive characterization, but it does capture at least some core elements of the notion and will suffice for current purposes.

In this chapter, I propose to discuss five different kinds of order – the natural order, the moral order, the political order, the divine order and the human order – and to consider (i) how fundamental shifts have occurred over the past several centuries that involve these kinds of order and (ii) what implications that has for our current understanding of the world and our place in it. More specifically, I will consider how each of these kinds of order deploys a particular notion of law, since it has traditionally been maintained not only that there are crucial dependency relations between these different orders, but also that a particular conception of law is at the heart of these orders and their dependencies. What will emerge from the discussion is that our current attitude towards these different kinds of order and their inter-relations is neither inevitable nor obviously coherent. As a result, these reflections offer us new perspectives on our present situation and on the possibilities that lie before us.

Modern conceptions of law in the natural, moral and political orders

The idea of a law of nature is very familiar to us today, given that discovering the laws of nature is one prominent goal of scientific inquiry. However, when one takes a broader historical view of things, it is quite striking that the idea of a law of nature was hardly mentioned at all in the natural philosophies of the ancient and medieval worlds and became prominent only in the early modern period with the onset of the Scientific Revolution. The idea originally took shape in the attempt to discover laws of motion in particular, which would describe the behaviour of a specific range of inanimate bodies, but given the success that leaders of the Scientific Revolution had in discovering such laws in the realm of mechanics (culminating in Newton's *Principia*), it was no great surprise when they did not rest content with providing explanations of inanimate objects, but expanded the scope of their search so as to encompass the laws of nature in general. Specifically, proponents of the Enlightenment were especially keen on the attempt to discover laws that would govern the behaviour of human beings, which would form an integral part of a science of man that would run parallel to the science of nature. Indeed, such laws would be essential to making good on the Enlightenment's call for social progress, especially among the emerging middle class, whose dramatic gains in literacy and power during the eighteenth century made possible the fruitful application of scientific knowledge to both their professional and personal lives.

Though early modern natural philosophers offered a variety of proposals regarding the proper formulation of the laws of motion in particular and of the laws of nature in general, there was widespread agreement about the core notion of a law of nature. The basic idea was that a law of nature is a principle that would govern what would happen in nature by determining what the next state of the world would be in light of the previous state of the world. The scope of a law of nature was supposed to be quite broad so as to encompass any thing of a given type, regardless of the particular state that it happened to be in. Thus, the laws of motion were supposed to govern the behaviour of *all* inanimate bodies, regardless of how fast or slow they might be moving and in what direction. As a result, this meant that a law of nature was often explicitly understood as an exceptionless regularity for a particular kind of object. It was also typically presupposed that laws governing bodies must be describable in mathematical terms, since that allowed for a much higher degree of clarity and precision in calculations and predictions than previous accounts were capable of. Given this conception of a law of nature, the laws of nature are fundamental to what happens in the world and, as a result, the overall task of natural science was to formulate the smallest number of laws that would explain the largest number of phenomena. The extraordinary status of Newton's law of universal gravitation in science derived from its extraordinary success at this task, since it was a *single* law that determined *all* motions of both celestial and terrestrial bodies with an unsurpassed degree of *mathematical exactness*.

Though laws of nature were thus considered fundamental to scientific explanations of natural events, almost no one in the early modern period thought that the laws were *absolutely* fundamental. It was widely held that the laws of nature depended on God in some way. It is important to note that the basic motivation for claiming that the laws of nature depend on God was not necessarily or exclusively theological, even if many at the time thought it obvious that Scripture should be read as entailing the dependence of everything, including laws of nature, on God. The strictly philosophical line of thought runs as follows. If one asks why a certain event has occurred in nature and is told that it followed from a law of nature, the next question is inevitably why that particular law of nature exists. Now it sometimes happens that one law of nature depends on another (e.g. by being a more specific instance of it), but once one has reached the basic laws of nature, it is clear that nothing within the natural world could explain why these most basic laws exist. After all, the most basic laws are supposed to explain the events that occur in nature, not be explained by them. At the same time, even these most basic laws have particular features that are such that one can, at least in principle, imagine other possibilities. (For example, we can speculate about what the world would be like if Planck's constant had been twice as great or if $F=ma^2$ instead of $F=ma$.) And if one can imagine other possibilities, then it seems appropriate to call for an explanation of why these laws exist, and not any others. At this point, if one is already committed to the existence of God on other grounds, as many early modern natural philosophers were, the most expedient explanation would be to assert that the laws of nature must be grounded in God.

Now the question of how exactly the laws of nature could or should depend on God was hotly contested at the time, with three main kinds of answer vying for acceptance. One option, articulated by René Descartes, is as follows. If God is an absolutely perfect being, that is, a being endowed with all possible perfections, including omnipotence, omniscience and omnibenevolence, it is plausible to think that God must be immutable. If God were to change, that change would have to be either from better to worse or from worse to better, but neither of these options is compatible with God's absolute perfection, since God is supposed to be perfect at all times. But if God is immutable, then the way in which he creates and sustains the world in existence from one moment to the next cannot change either. Insofar as the laws of motion are principles that describe how the world changes from one moment to the next, they are simply descriptions of the immutable ways in which God creates and sustains the world. As a result, the laws of nature can seem to follow from God's immutability. This line of thought finds support in the fact that Descartes's laws of motion include a conservation law, which asserts that the same quantity of motion must be preserved throughout all changes.

A second option, developed in detail by Nicolas Malebranche, a priest in Paris who was otherwise sympathetic to Descartes, is that the laws of nature derive from God's simplicity. If it is better to bring something about in a simple rather than in a complicated way, it is clear that simplicity is a perfection. If God has all perfections, God must have the perfection of simplicity. Now insofar as the laws of nature are supposed to be constituted by a small number of principles that have universal scope, it is clear that the laws of nature are consistent with the simplicity that pertains to the ways in which God creates the world and maintains it in existence. While one might think that God could repeatedly choose anew how the world should be at each moment in time so as to be able to make adjustments as needed, this would obviously be a complicated and potentially ad hoc decision process. It is both simpler and better reflective of God's infinite wisdom to have God decide once and for all on a small number of general policies that exist for all cases at all moments in time. Insofar as the laws of nature are the simplest possible policies by which God determines what happens in the world at each and every moment in time, it is plausible to maintain, Malebranche thinks, that the laws of nature derive from God's simplicity and wisdom.

A third option, described by Gottfried Leibniz, who was a fierce opponent of Newton, is that the laws of nature derive from God's omnibenevolence. If one focuses on God's goodness, then it is clear that the world that God creates must be the best. For if God could have created a world that is better than the one he did in fact create, he would have been an underachiever. But since God's omnipotence rules out the possibility that he did not have the power to create the

best world and his omniscience rules out that he might not have
known about all of the possibilities that he has the power to create,
it is clear that the deficiency that would result in the creation of a
sub-optimal world would have to lie in a lack of goodness. However,
since God's goodness is maximal, it is clear that God cannot be an
underachiever by creating a world that is less good than it could have
been. Now if the best world contains laws of nature, as our world
seems to, then God's goodness must be responsible for these laws,
especially insofar as it is obvious that a perfect being would act in an
orderly way. Although we can imagine lawless worlds, they would,
Leibniz thinks, be disorderly disasters of random events, nowhere
near as good as the orderly world God decided to create due to his
benevolence.

Despite the interesting differences between these three conceptions of how the laws of nature might depend on God, the similarities
are striking and important. The most important similarity is that
whether the laws derive from God's immutability, his simplicity or
his goodness, they all depend on characterizing God as a perfect
being, since these features are either perfections or immediate consequences of perfections. Further, insofar as God uses the laws of
nature to impose order on what happens in nature, the laws of nature
are crucial to defining what the order of nature is. That is, whatever
happens according to the laws of nature is *eo ipso* a natural event and
part of the regular order of nature. The flip side of this is that any
exceptions to the laws of nature are irregularities, deviations from
the order of nature that require special explanation. Thus, instead of
thinking of apparent exceptions to the natural order as monsters or
freaks of nature that do not admit of explanation, the early moderns
viewed such events either as following from laws of nature in ways
that we do not yet fully understand, or else as miracles – that is,
supernatural events that were caused directly by God for special
reasons. Many early modern thinkers held that the special reasons

that one might cite to explain the occurrence of miracles were ultimately founded on moral reasons belonging to the order of grace, which pertains to God's decision as to whether human beings should receive salvation. One way of describing this, which Malebranche preferred, is to say that miracles are possible, on this account, because the order of nature as determined by God's wisdom is subordinate to the order of grace, which reflects God's ultimate purpose in creation. Yet the main point that the early modern treatment of the laws of nature brings out quite clearly is that the natural order, as defined by the laws of nature, as well as any deviations from it depend on a more primitive divine order that is ultimately based, in one way or another, on God's perfection.

Although the idea of a law of nature became prominent only in the early modern period, the idea of law itself was of course much older, occupying an important place in both political and moral contexts in the ancient and medieval worlds. Thus, political authorities, whether monarchs, consuls or other legislative bodies, would promulgate and enforce political laws that bound their subjects, or citizens, while the moral law obligated all human beings alike, regardless of their social and political status and circumstances. Though kings and queens often exempted themselves, everyone else, at least in principle, was subject to whatever political laws were enacted, and the moral law applied universally, even to the most powerful of royalty, so that both political and moral laws were either exceptionless or nearly so. In this regard, they displayed precisely those features that were fundamental to laws of nature, which derived from this older conception of law, albeit with important modifications.

Now what stands out in the current context about the idea of moral and political laws is that early modern thinkers followed their medieval predecessors in maintaining that these laws depended on God in ways that are analogous to the way that laws of nature depend on God. Again, though there were significant disagreements on the exact nature of the dependence relation, both political and moral laws were thought to require a lawgiver who is superior to those who are subject to the law in question, who issues threats of punishment and promises of rewards and who is capable of providing sufficient motivation for their compliance with the law. It is this conception that gave rise to the slogan: No law without a lawgiver. The major point of contention was thus not whether a moral or political law depends on God, but rather whether the obligation to perform an action depends solely on some kind of reason that would be equally intelligible to both God and man, as intellectualists claimed, or whether it depends solely on the (to us unfathomable) will of God and is expressive of the kind of divine command theory espoused by voluntarists, according to which one should do an action simply because God commanded it. Thus, early modern thinkers held that moral and political laws, which define what the moral and political orders are, depend on God in a way that parallels the dependence of laws of nature on God. As a result, despite all of their differences, the early modern philosophers offer a comprehensive and powerful world view: the moral, political, and natural orders are all defined by (nearly) exceptionless laws that depend on the divine order, which is determined by some features of God's perfection.

This early modern conception of law, and of the natural, moral and political orders that were thought to depend on law, was modified in different ways as the modern period progressed, but amidst all the differences a common theme emerged over time, namely the invocation of the human order. For example, David Hume, a Scottish empiricist, attempted to redefine the natural and moral orders by way of appealing to the psychological capacities of human beings. That is, instead of thinking that the natural and moral orders reflected an independent order in the objective world, Hume reduced the order we find in nature to our psychological dispositions. Thus, laws of nature are really based on our own psychological expectation that the future will behave as it has in the past, and the moral law can be reduced to a purely subjective, albeit regular feeling of sympathy that arises when human beings come into contact with each other. Now Immanuel Kant, a Prussian through and through, famously rejected Hume's position, since he was dissatisfied with, among other things, the contingency inherent in the subjective psychological capacities that Hume invoked. Instead, Kant claimed that both the laws of nature and the moral law derived from our rationality (rather than from any empirical psychological disposition that we might or might not have or that might change over time) such that we legislate not only moral and political laws, but also the lawfulness of the laws of nature. But, despite these quite fundamental differences, Kant shared with Hume the idea that the natural and moral orders could be captured in terms of laws for which human beings were responsible. In short, the natural and moral orders depend, for both of these thinkers, on the human order (even as they disagreed about what precisely was essential to the human order). While it might seem to be an unimportant addendum to relate the natural and moral orders to the human order in this way, one of Kant's great achievements was to show how two of our most fundamental values in the Western tradition, freedom and autonomy (which is self-legislation), were possible only if one accepted the centrality of the human order to the natural and moral orders.

The picture that thus emerges from our brief consideration of modern conceptions of the notions of law and order is one in which the natural, moral and political orders are defined in terms of a substantive notion of law that depends on either a divine order or a human order. For both the major advances in science and the moral and political reality of the day could not be made sense of during this period without the notion of law, and the notion of law, in turn, requires some kind of lawgiver, whether such a legislator be divine or human. Such were the most basic concepts and claims with which people attempted to make sense of the world during the modern period.

Contemporary conceptions of law and order

When we now compare this modern conception of law and order to our contemporary conception, we immediately see that while some features of the early modern conception have remained largely intact, others have been either modified in some respects or rejected altogether. The most significant constant here is the notion of a law which is still present both in many of the natural sciences and in moral and political contexts. However, two significant differences are immediately apparent. First, the connection between the natural order and laws of nature has been loosened at least to a certain degree, since it is clear that not all natural sciences are directed at discovering universal laws. Biology is not as wedded to discovering laws as mathematical physics has been and even physics may not hold that the laws of nature have as universal a scope as had sometimes been thought.

A second significant difference is that it has become much less popular to maintain that law, in any of its forms, necessarily depends on God. This is most obviously the case in political contexts, at least in Western Europe and North America, where theocratic forms of government have given way to democracies in all but name, but it is standard in the moral realm as well insofar as the idea of a purely secular moral law is widely (though not universally) accepted, even if morality is often privately associated with the idea of divine commandments (and their future heavenly rewards and hellish punishments). Though both Newton and Einstein thought that discovering the laws of nature was akin to divining God's intentions – e.g. Einstein remarked, against Bohr, that 'God does not play dice' to explain his rejection of indeterminacy – it is clear that science as practiced today does not depend on widespread and explicit appeals to religious claims of any sort. So it is much less popular to continue to maintain the dependence of the natural, political and moral orders on the divine order.

In the context of this second difference it is interesting to note that claims to the existence and explanatory relevance of God had already come under attack in the modern period. Though various lines of criticism were developed, one line, articulated especially forcefully by Hume, questioned whether the natural order is at all attuned to the moral order, or is rather entirely impersonal, with the implication being that if the natural order were impersonal, it could not depend on a benevolent God. Thus, if our evidence for God's existence must derive from those features of the natural order that make it seem as if the world was created for our benefit, and if we deny that the natural order has such features, then we will never be able to infer God's existence, much less any possible implications for our future lives. Further, accepting that the natural order is thoroughly impersonal makes it more difficult to maintain that the moral and political orders depend on God as well. In one fell swoop, following this line of thought, the necessity of the divine order is seemingly undercut.

Now one reaction to these historical developments is essentially conservative. We once had a conception of order such that the natural, moral and political orders depend on God and the divine order, and we now have one in which that is no longer the case, but, so the conservative impulse maintains, everything can still proceed as it was, basically unchanged. That is, although there is no divine order, there is also no need to replace it with anything else, since we can continue to adhere to the natural, moral and political orders just as before by continuing to seek laws of nature in science and to act according to established moral and political laws. However, such a conservative approach may not ultimately be coherent. For it is not

obvious that the notion of law is instantiated in the world (as opposed
to being an admittedly useful figment of our imagination), and if we
have no reason to accept that the world is or ought to be ordered
according to laws in the guise of exceptionless regularities, we would
have a substantive hole right in the middle of our conception of the
world. Without the notion of a law, what keeps our world from being
a place of unpredictable turmoil and disorder, where things happen
one way some of the time, but in different ways at others? Further,
without a divine guarantee of their consistency, we would have no
reason to assume that the laws of nature are compatible with the
moral and political laws. As a result, this option – seeking laws with
no justification for assuming their existence – seems to be relatively
unattractive, even though much contemporary thought implicitly
assumes a model of roughly this kind.

Faced with this situation, one could instead follow Hume or
Kant by claiming that the natural, moral and political orders can be
justified by means of the human order. As we saw above, Hume bases
our belief in the lawfulness of the natural and moral orders upon
our propensity to form habits (whether they concern inferences or
feelings of sympathy) that are fundamentally psychological in kind,
and Kant justifies natural, moral and political law by way of reason
and its legislative capacity (which is supplemented with a 'tribunal
of reason' to adjudicate conflicting claims). Thus both maintain that
the human order in effect replaces the divine order as the immediate
foundation for the natural, moral and political orders. It is important
to note, however, that in Kant's case, the divine order does not drop
out entirely, since he thinks that the moral law presupposes not only
our freedom, but also belief in God's existence and the immortality
of the soul. As a result, it is more accurate to say that, for Kant,
the human order is a (crucial) mediating link between the natural,
political and moral orders, on the one hand, and the divine order, on
the other. How the details of such a justification are to be worked out

and whether it can succeed have been topics of debate, but one can immediately see the attractions of such an approach, which departs from the conservative reaction just described in significant ways.

Another option, which could still be considered relatively conservative (at least compared to the broad spectrum of possibilities), would be to retain the idea that the natural, political and moral orders presuppose the divine order, just as orthodox Christian theism maintains, and to explain how it is that contemporary scientists have missed the dependence of the worldly orders on God. Many strategies can be and have been pursued, including the idea that science, with its distinctive methodology, is well-suited to the investigation of the natural order, but oblivious, perhaps necessarily so, to the supernatural order that is responsible for it, which other disciplines, such as theology and philosophy, are in a better position to investigate. That is, perhaps science, philosophy and theology are different disciplines with different methods and subject matters, with the result that science, with its orientation towards the world, is not in a position to understand itself and, as a result, it fails to see the limitations of scope that are entailed by its method.

There are, however, other, more radical options. One reaction, with historical precedents among Protestant theologians, such as Friedrich Schleiermacher, is to redefine the divine order such that it no longer conflicts with the natural order. For example, one might think that the problems of the early modern conceptions arise from assuming that God is a simple, immutable being that, due to a certain conception of his perfection, is constrained to act according to universal rules that then take the form of impersonal laws of nature. If one rejects this assumption, as Keith Ward argues one should, one can instead appeal to an essentially personal God, who reacts and responds to human beings and their plights. Such a conception allows for a different understanding of God's relation to the natural and human orders. Again, many details would need to be worked out, but the historical precedents (in the ancient, medieval and especially modern world, with Schleiermacher and other theologians) are a useful resource in engaging in such a project today. A new conception of the natural order can then make possible a new conception of natural theology, in particular, new possible theistic proofs, which Russell ReManning explores.

However, there are further interesting options which are more radical still. Instead of either attempting to show how the human order can replace or supplement the divine order or redefining the divine order, one could reconceive of the natural order by denying that laws, understood as exceptionless regularities, are required to define the natural, moral and political orders. Contemporary approaches in the philosophy of science have pursued this option in a number of interesting ways. For example, one line of thought, which Nancy Cartwright has emphasized and Steven Horst articulates in one particular direction, focuses on the notion of a capacity, or causal power, instead of a causal law (understood in a particular way). A second idea, which Roman Frigg and Robert Bishop articulate, is to emphasize the notion of self-organization. One more specific way of developing the notion of self-organization derives from the notion of free agency and the special way in which such agents can act in the world, as illustrated by Tim Mawson. A fourth idea, which gained widespread popularity in the 1980s (due to work by Alasdair MacIntyre), has been to understand morality not in terms of moral laws, but instead as consisting of the virtues, just as Aristotle had emphasized long ago. A fifth idea, pursued by Harre and Montuschi, attempts to redefine the social order independently of the natural order. Of course, it is also possible to combine several of these approaches, as Bechtel and Martin have done when emphasizing that defining a notion of the natural, biological order could proceed by invoking taxonomies, mechanistic explanations and evolutionary patterns of descent with modification.

Though the project of redefining the natural order altogether might seem to suggest that the historical conceptions described above are best left behind, it turns out that such conceptions could still be of considerable use at this point, since several of these 'new' definitions have important historical precedents, which can provide a terminology and conceptual framework with which one can work out the new definitions. For example, appealing to causal powers or capacities, as Cartwright does, is precisely the approach that was invoked in natural philosophy during the ancient and medieval periods by Aristotle and his followers. In their reflections on how to account for the natural order, many ancient and medieval Aristotelians focused on the natures of things and used an understanding of those natures in considering how they could and would relate to each other, where the natures are not immediately identified with laws of nature (even if they might be related in some way). Further, the idea of free agency has, throughout much of the history of philosophy, been viewed as requiring a special status for persons, since free will requires persons and persons are distinct in kind from that of the physical states of inanimate bodies. At the same time, the notions of a person and of the free agency that goes along with it fit in quite naturally with an Aristotelian ontology of qualitatively distinct natures and causal powers that act according to them. Finally, the notion of self-organization and the special kind of causal features that are required for self-organization have been emphasized by Kant, who provided a sophisticated analysis of how self-organizing entities relate to the mechanistic laws established in physics.

As a result, there is a rich storehouse of concepts and ideas that one could appeal to in undertaking the task of redefining the natural order; in this way (and in many others), the history of philosophy can be of considerable use as we move forward today. Further, by recognizing that the seemingly radical options that we confront today have important historical precedents from which we can still learn, we can also see that there is a genuine sense in which these radical options are not, in the end, so radical after all, but rather represent a return to more traditional ideas, and the early modern position can seem like a very brief interlude that interrupted the long course of a different, more traditional approach.

Conclusion

What these historical reflections show is that we face a deep and fundamental challenge today. The most general structure that established an intelligible relation between the various orders of our world – natural, moral, political, human and divine – has collapsed and we must either find an appropriate replacement for it or else learn to live with an uncomfortable amount of intellectual disorder and disarray. I have indicated what some of the possibilities are, as well as what tasks lie before each option, so I do not believe that our situation is hopeless, but it is, I think, clear that important tasks still lie ahead of us.

2

The Dethronement of Laws in Science

Nancy Cartwright

Editorial Link: Nancy Cartwright is perhaps the best-known proponent of the new view of order in nature, which she calls a 'dappled world', a world filled with diverse sorts of causal principles interacting in various ways in different historically developing contexts. She argues that not everything in nature is precise, measurable or mathematically expressible. Experimental practice in the sciences is often pragmatic or based on intuitions and uses all sorts of creative and 'clever engineering' and a 'motley assembly' of varied tools and techniques to solve practical problems. Theories can be idealized and simplified models of real-world interactions, which are more fuzzy and complex. New principles come into play at new levels of organization. Forms of order can be local and fragile, both dependent upon the contexts within which they exist, and susceptible to modification by many different forces. Experimental scientists do not need to postulate one exceptionless set of universal principles. They work with local, piecemeal, contextual and developmental schemes of order. In this sense, there is 'order without laws'. KW

From the faceless particles of fundamental physics to marshes, mountains and rain forests, fleas, walruses and traffic jams, we used to live in a world governed by eternal, all-encompassing laws, laws discovered by the experiments of physics and encoded in its mathematical equations. At one time these laws were laid down by God, who commanded all His creations to act in accordance with them, without exception or excuse. After a while God faded from the world that science pictured – it seemed nature needed only the laws to know how to carry on, relentlessly, predictably. Now the laws are fading too.

This 400-year-old image of the governance of nature by eternal, universal laws is today being undermined by exciting new modes of understanding across the sciences, including physics and biology, as well as, perhaps less surprisingly, in the study of society. There is order visible in the world, and invisible. But if we trust to these new ways of understanding, this need not be order by universal law. It can be local, piecemeal and contextual – much like the world as we see it around us.

We live our everyday lives in a dappled world quite unlike the world of fundamental particles regimented into kinds, each just like the one beside it, mindlessly marching exactly as has forever been destined. The everyday world is one where the future is open, little is certain and the unexpected intrudes into the best laid plans, where everything is different from everything else, where things change and develop, where different systems built in different ways give rise to different patterns.

Philosophers call the reflective picture of what we experience in our everyday lives the *manifest image*: 'manifest', as the *Concise Oxford Dictionary* says, 'shown plainly to the eye or mind'. The *philosophical* contrast is with the *scientific image*, the image of the world as our best science pictures it. For centuries since the Scientific Revolution, the two have been at odds. But many of the ways we do science today bring the scientific and the manifest images into greater harmony. For much of science understands and models what was previously unintelligible; it devises methods to predict what will happen in the future (and with very great precision) and builds new technologies to manipulate the world around us without resort to universal laws.

This breakdown of universal law is not along any one fissure nor provoked by any one great discovery like the quantum of action, nor does it emerge from just one science. Rather, it appears in many distinct, highly detailed studies of scientific practice. Though generally unrelated to one another, these diverse studies have in common a radical split from the standard view. They propose alternatives to universal laws as the central explanatory and predictive mechanisms employed in the sciences. This questioning of the order of science has come from analyses of successful scientific practice across the disciplines, from fundamental physics through biology to political economy.

The reaction against the received view of scientific laws as universal and exceptionless has been as diverse and widespread as the scientific disciplines from which criticisms have been developed. No one view is the same as any other. But they have in common a picture in which laws are not the immutable and exceptionless governors of a nature completely ordered under them. In the next section of this book you will find detailed studies from the natural and social sciences. Here I provide a brief overview of some of the recent more general work exploring the dethronement of universal laws in the two natural sciences discussed there: biology and physics.

Biology

Biology has long been criticized: It is not a proper science because it does not have proper laws. Now those who study its various practices and the many impressive successes they produce are fighting back. Perhaps the traditional view of what counts as proper science with proper laws has been mistaken all along. Contemporary biology seems to have just what it takes to describe nature successfully and to put its knowledge to use.

One of the most vocal scholars arguing that this is in general true across biology and even far more widely is University of Pittsburgh philosopher of biology, Sandra Mitchell:

Take … Mendel's law of segregation. That law says in all sexually
reproducing organisms, during gamete formation each member of
an allelic pair separates from the other member to form the genetic
constitution of an individual gamete. So, there is a 50:50 ratio of
alleles in the mass of the gametes. In fact, Mendel's law does not hold
universally. We know two unruly facts about this causal structure.
First, this rule applied only after the evolution of sexually reproducing
organisms, an evolutionary event that, in some sense, need not have
occurred. Second, some sexually reproducing organisms don't follow
the rule because they experience meiotic drive … Does this mean
[that] Mendel's law of segregation is not a "law"?[1]

From hosts of cases in biology like this various authors conclude that,
rather than good old-fashioned 'proper laws', biology offers instead:

• Laws that emerge historically.
• Laws that are contingent.

These two come naturally from Mitchell's first two sources of
unruliness. They also conclude that biology offers only

• Laws that are not exceptionless.

This is a natural account given Mitchell's second source of unruliness.
    Different kinds of cases far from evolutionary biology, in molecular
or neurobiology for instance, lead others to propose that biology studies
not laws that describe regular behaviour that must occur but rather

• Mechanisms that, functioning properly, in the right places,
  can generate regular behaviour, for instance the interactions of
  the structures of non-RNA strands with tRNA molecules and
  ribosomes to underwrite protein synthesis.

Mechanisms are a central topic of our chapter on biology in Part 2
of this book.
    Mitchell also points to authors who claim that biology studies

- *Ceteris paribus* laws, laws that hold only in special circumstances.

She herself advocates something more practical and this is a piece of advice that she proposes to carry outside of biology across the disciplines, from economics to physics:

> We need to rethink the idea of a scientific law pragmatically or functionally, that is, in terms of what scientific laws let us do rather than in terms of some ideal of a law by which to judge the inadequacies of the more common (and very useful) truths [of the kind biology teaches].[2]

Mitchell argues that we should do away with the old dichotomy 'law vs. non-law' or what is universal, exceptionless, immutable versus all the rest, to be replaced by a sliding scale, and along a variety of different dimensions:

> [All] general truths we discover about the world vary with respect to their degree of contingency on the conditions upon which the relationships described depend. Indeed, it is true that most of the fundamental laws of physics are more generally applicable, i.e. are more stable over changes in context, in space and time, than are the causal relations we discover to hold in the biological world. They are closer to the ideal causal connections that we choose to call 'laws'. Yet, few of even these can escape the need for the ceteris paribus clause to render them logically true.[3]

Looking at how the successes of science are produced across the disciplines, it is truths of varying forms with varying degrees of universality and exceptionlessness, describing various degrees and kinds of order, that let us do what we need to do to produce those successes.

Emergence and vertical reduction

Mitchell's final remarks bring us to physics, which is eventually where the conversation goes when order is challenged: 'Surely the world of physics is totally ordered and if physics is ordered, so is all the rest.' Notice that there are two claims here. The first is that the world of physics is totally ordered. The second is that physics fixes all the rest. This second doctrine usually goes under the title 'reductionism' but we call it 'vertical reductionism'. (You will see why in a moment.)

The opposite of vertical reduction is generally labelled 'emergence': the idea that the more 'basic' levels of reality do not determine or fix what happens at 'higher' levels; that new phenomena, new characteristics, even new laws of nature emerge at larger dimensions, more mass, higher velocities or increased complexity. There has been a great deal of work on emergence recently and there are many non-technical accounts available of the central ideas and issues so I will comment only briefly on it before turning to our central topic for physics and to our topic of focus there, what might be thought of as *horizontal reduction*: the idea that the cover of natural law, at any one level or crossing all levels, may not be complete; that order may remain to be made in nature by us, not just discovered.

In philosophy of science, vertical reductionism probably reached its zenith in the early 1960s, in the wake of Ernst Nagel's *The Structure of Science*, which attempted a rigorous reconstruction of classical mechanics. But those who attempted to carry on Nagel's project outside classical mechanics soon found that classical mechanics was, at best, a special case where reductive strategies came close to working. And indeed they did not ever truly meet the standards of demonstration even there.

Here is a partial list of problems:[4]

- Often important features of the macrosystem were left unexplained by the supposed reduction base. For example, the supposed reduction of the ideal gas laws to the collisions of particles in statistical mechanics is often held up as a paradigmatic example of successful reduction. But even here, a crucial thermodynamic property – the direction of entropy – is left unexplained. Statistical mechanics is temporally symmetrical, thermodynamics is not.

- In some sciences, the classification of objects uses functional principles – say, reproduction and nutrition in biology, or organs in anatomy. It may well be that a micro explanation of each individual organism can explain why a certain process works as reproduction or nutrition, or a particular organ functions as a heart. But human hearts are very different anatomically from earthworm hearts, and there is an even greater difference between the mechanisms of nutrition and reproduction of animals, plants, protista and fungi. And so what unites human and earthworm hearts into a single category, or digestion and photosynthesis into types of nutrition, is not explained by looking at the subsystems.

- Some scientific kinds, like biological species, have an irreducibly historic dimension. If there are beings just like us structurally on some distant world but not members of a common lineage, they are, from a biological standpoint – or at least one important way of understanding species within biology – different species.

- Macrosystems may also have properties that are not predictable from lower levels and may play an important top-down organizing role, as for instance in living systems, where there are important jobs for high-level regulative processes that exert top-down control.

- The categories and processes typical of a macrosystem may also need to be understood in terms of how that system is embedded in an even larger system, for example when we are describing a feedback system or engaged in ecological biology.

- And perhaps most fundamentally, there are multiple ways of decomposing a macrosystem into parts. Consider the case of an organism – say the various ways one needs to understand the workings of a human body in order to have a grasp of modern medicine. These include of course a way of decomposing the organism into more basic anatomical components. But there are also biological 'systems' that do not break down across anatomical lines, but are distributed all over the organism, like nutrition and the immune system. Gene expression also takes place everywhere in the organism. And while the endocrine system has a few anatomical organs that produce hormones, the activity of these likewise takes place throughout the organism. Moreover, medicine is very much in the business of classifying the functioning of an organism as normal and abnormal in various ways, at least some of which require appeal to the organism's relationship to its environment and in some cases to the ancestral environment for which a system is adapted.

Physics and horizontal reduction

Let us turn now to horizontal reduction: Does physics offer laws that dictate a complete order among the very kinds of events that physics studies? Before looking at reasons in recent studies for thinking not, it is good to make sure the question is clear. One usual question debated by philosophers and physicists alike is

Realism: Are the well-confirmed laws of physics likely to be 'approximately' true?

This book on the whole takes a realist stance: Their remarkable success at precise prediction and at new technologies gives us warrant to suppose that our modern sciences are getting at the truth. This is controversial in some quarters. We propose to sidestep this controversy because the horizontal reduction we focus on is about a very different question, about the dominion of physics:

> *Dominion:* Does physics dictate everything that happens in the physical world?

This question is not about whether the laws of physics are true but about how much territory they govern. A number of different kinds of considerations about the way physics functions when it functions best at predicting and intervening into the world around us argue that the answer is no.

Ceteris paribus laws

The first suggests that, just like the laws of biology as Mitchell sees them, the laws of physics are *ceteris paribus* laws: laws that come with the caveat that they hold so long as 'other things are equal' or, closer to what is really intended, 'in particular circumstances'.

Consider philosopher of physics Marc Lange:

> The law of thermal expansion must also include a great many qualifications. If a complete, explicit statement of the law must specify each of the restrictions individually, then the statement will be very long indeed. It must specify not only that no one is hammering the bar inward at one end, but also, for instance, that the bar is not encased on four of its six sides in a rigid material that fails to yield as the bar is heated. For that matter, it should not deem the law inapplicable to every case in which the bar is being hammered on. The bar may be hammered on so lightly, and be on such a frictionless surface, that the hammering produces translation rather than compression of the bar ...

However, as a mere practical difficulty, this does not seem terribly serious. It can easily be avoided by employing some such qualification as "in the absence of disturbing factors," "ceteris paribus," or "as long as there are no other relevant influences." I term these sorts of qualifications *provisos*. But once we set the practical difficulty aside in this way, we are confronted by a more fundamental problem. What does a proviso mean?[5]

Lange is considering a canonical reply by defenders of the universal rule of physics. The idea is that the list of restrictions that must be specified

... is indefinite only if expressed in a language that purportedly avoids terminology from physics. If one helps oneself to technical terms from physics, the condition is easily stated: The "law" of thermal expansion is rigorously true if there are no external boundary stresses on the bar throughout the process.[6]

This brings into focus the technical terms of physics, like 'stress'. How far do these stretch? Here the very virtues of physics get it into trouble. The terminology of physics is tightly controlled. This is what distinguishes it from disciplines that hardly count as science at all, that use terminology like 'globalization' or 'unconscious desire', terms that have no such rigid criteria for their application. There are rules in physics for how to use language, how, for instance, to ascribe a quantum field or a classical force: rules like $F = GMm/r^2$ *when* two masses a distance **r** apart interact. And in most situations there are a number of factors affecting the outcome that we do not know how to describe using these regimented descriptions. Yes, the law of thermal expansion may be exactly true if the restriction 'and there are no external boundary stresses' is added. But will it continue to be rigorous? Is 'no external boundary stresses' a proper scientific description with precise rules for its application, or is it rather, as defenders of *ceteris paribus* laws argue, a loose catch-all term that marks that the law is not universal and that there is no universal law using proper physics concepts that can stand in its stead?

Consider the Stanford Gravity Probe Experiment, on which I was for a while a participant observer. After over two decades in development the Gravity Probe put four gyroscopes into space to test the prediction of the general theory of relativity that the gyroscopes would precess due to coupling with space-time curvature. The Stanford Gravity Probe prediction about its gyroscopes was about as condition-free as any claim in physics about the real world could be. That's because the experimenters spent a vast amount of time on the project – over 20 years – and exploited a vast amount of knowledge from across physics and engineering. The experimenters tried to fix it so that all other causes of precession were missing; hence all the other causes would be, *ipso facto, describable in the language of physics!* Moreover if they had not succeeded and other causes occurred, then *any that they couldn't describe would make precise prediction impossible.* If you can't describe it, you can't put it into your equations.

It should be no surprise then that all the good confirmations of the laws of physics occur in the special situations where we can describe all the causes with proper physics concepts. That is the real content of the 'ceteris paribus' clause: This law holds so long as all the factors that affect the behaviours under study can be described by proper physics concepts that have the kind of strict rules for their application that good, rigorous science demands; it holds only in specially structured environments where we can describe all the causes at work – indeed these are the only environments where we can produce such predictions. Whether there is systematicity outside environments structured like this is speculation. So too is the assumption that all environments are secretly structured in the right way, even if we have not yet discovered it.

Laboratories are structured in the right way, and lasers, batteries and bicycles. So too are a very great many naturally occurring situations. The planetary system is so structured and seems to have little disturbance that cannot be subsumed under proper physics concepts.

But most situations do not seem to be structured in the right way. In physics there is no rule that takes you from 'the bar is being hammered' to 'the bar is subject to certain stresses'. The rules for assigning terms like 'stress' require not loose terms of everyday use to apply but a far more technical, regimented vocabulary. Stresses and strains are characterized as forces and force functions, and as I already noted have strict rules for application, rules like $F = GMm/\mathbf{r}^2$ *when* two masses $\mathbf{r}$ apart interact, $f = \hat{I}_0 q_1 q_2/\mathbf{r}^2$ *when* two charges interact, and so forth. And it is not clear that every time a bar is hammered one of these more technical descriptions obtains. Physics is, above all, an exact science. Its concepts must be precise, measurable and fit in exact mathematical laws. This means that they may well not be able to describe everything that affects even outcomes reasonably supposed to be in physics' own domain.

Those who argue that the laws of physics are for the most part *ceteris paribus*, holding in special circumstances rather than universally, are generally great admirers of physics. There is no implication in this suggestion that these laws are not getting at the truth. You can be a realist about *ceteris paribus* laws, and for the same reasons that are often alleged in support of the approximate truth of universal laws: the power the laws of physics – now seen as *ceteris paribus* laws – exhibit to produce precise predictions and to help us build fabulous new technologies to intervene in the world and to change it.

In defense of the universality of the laws one can of course *say* that the bar *must be* subject to a stress because of the way its behaviour is affected. But that reduces physics to the status of psychoanalysis. We can *say* that I have certain unconscious desires because of the odd way my behaviour is affected. A physics that allows us to say things on that kind of basis is not the physics that yields the precise predictions and exact control of nature that gives us reason to think its laws are true. Better to suppose the laws are *ceteris paribus* than to deprive them of their power to predict precisely and of the huge empirical support this power provides.

Incommensurability

A second kind of study arguing against horizontal reduction looks at the interactions among different kinds of physicists, for example instrument physicists, experimentalists and theorists. Much of this work comes from Harvard historian of physics Peter Galison, beginning with his studies of neutral currents and the hunt for the Higgs particle. This particle plays a central role in the standard model of particle mechanics and in the grand unified theory. It was first predicted fifty years ago and was finally detected in the Large Hadron Collider in CERN in 2012. Galison's original studies were on the relation of theory to experiment in the 1970s but he has continued work on the issues ever since. Galison points out that much of the time experimentalists and theoreticians use many of the same words but mean something very different by them. That's because the meaning of the technical terms in physics is given not by single definitions in language antecedently understood but rather by the whole network of assumptions and inferences with other technical terms that can be made using them. Experimentalists and theoreticians make very different assumptions and inferences, caring little about the bulk of what the other says about neutral currents or the Higgs particle and often not understanding it. This is like the well-known incommensurability thesis of Thomas Kuhn: that the meanings of the terms used in physics depend intimately on the whole web of assumptions in the theory that use those terms and relate them to others. So, although two different theories may use the same sounding term and seem to make opposing claims using it, they cannot really contradict one another because the term does not mean the same thing in the two different theories. Galison's thick descriptions of the kinds of different assumptions that the experimentalists and theoreticians make about neutral currents or the Higgs particle flesh out this thesis for a real case in contemporary physics.

What Galison contributes that is really new is his account of how the two groups do communicate, as they do when theoretical claims are tested. The two groups meet in what is analogous to a 'trading zone', where neither's home language is understood by the other. They use between them a trading language – a highly specific linguistic structure 'that fits between the two'. Really a pidgin or perhaps even a Creole dialect. Galison stresses that he intends 'the pidginization and Creolization of scientific language to be treated seriously just the way one would address such issues for language more generally',[7] not as a loose metaphor.

This is a nice solution to the incommensurability problem that rings true to the historical descriptions Galison provides. But pidginization raises real problems for the traditional account of laws. The laws of physics are supposed to be universal and comprehensive, but they are also supposed to be well tested. So which laws are these? It is the theoreticians' laws, fitting together into an integrated theoretical package, that have a claim to be comprehensive; it is the experimental laws with their connections to all the requisite experimentalist assumptions that are tested. And the pidgin laws will generally not have enough back-up from the home languages to do either. There is a plurality of laws here and together they serve our pragmatic needs, as Sandra Mitchell urges. Picking, choosing and combining, we can do what we need to do. But there is no set of well-tested laws that looks comprehensive enough to support total order in the domain of physics.

The motley assembly

A third kind of study that casts doubt on the possibility of horizontal reduction raises issues about the autocracy of physics. Much of my own work falls into this category. According to the *Concise Oxford*

Dictionary, autocracy is 'a system of government by one person with absolute power'. So

Autocracy: Is physics an autocrat within her own domain?

The answer should not depend on a quibble about the boundaries so let's focus on things that anyone would reasonably take to be in the domain of physics, such as lasers or high temperature supercon-ductors or the trajectories of the planets. Surely no-one would quarrel with the centrality of physics in building a laser, predicting the perihelion of Mercury or eventually understanding low temperature superconductors. But it is just these marvellous successes that bring into doubt the autocracy of physics. Even in her own domain physics does not reign on her own. She acts as part of a motley assembly.

Our treatment of the nucleus is a good illustration. Here we use a number of different models all at the same time. The problem of having many different and incompatible models for the same phenomenon is often seen as one of the biggest challenges for unified theories. It certainly raises big problems for horizontal reductionism.

First, these models are phenomenological in the sense that they are constructed not top down, by derivation from theoretical laws, but bottom up to account for experimental results. They use theoretical concepts, theoretical knowledge and theoretical insight, but piecemeal; they cannot be seen as what the supposedly universal theoretical laws dictate for systems of this sort.

Second, the theoretical concepts and theoretical insights employed come from different theories and sometimes these are theories that that are supposed to be inconsistent with each other – such as quantum and classical mechanics.

Third, radically different fundamental assumptions about the nature of the nucleus and its theoretical underpinnings are required to account for different experimental data. Different models account for different things. Sometimes we can see that these different models are idealized or simplified accounts geared to specific purposes but are all in fact, once the approximations and idealizations are understood, consistent with one consistent underlying account. The problem is that this is not always possible. So even here, where we are considering only abstract accounts of the nucleus and not real engagement in the world, we find we can't do the jobs we need to with one autocratic theory that rules the entire domain.

When it comes to detailed real-life prediction and intervention, autocracy fails even more dramatically. Let me illustrate from my own researches into the practise of physics. When I was at Stanford University I was in love with quantum physics and – being a committed empiricist – particularly with the startling empirical successes that speak for its credibility, especially lasers and super-conductors, which I made a special area of study. I was especially impressed simultaneously by how crucial quantum considerations are for understanding these devices and by how little they can do by themselves. They must be combined with huge amounts of classical physics, practical information, knowledge of materials and exceed-ingly careful and clever engineering before accurate predictions emerge, and none of this is described – or looks as if it is even in principle describ*able* – in the language of quantum physics. Physics can measure, predict and manipulate the world in precise detail. But the knowledge that produces our extraordinarily precise predictions and our astounding devices – the very knowledge that gives us confi-dence in the laws of physics – is not all written in the language of physics, let alone in one single language of physics. Its wellspring is what I call 'the scientific Babel'.

I was clearly influenced in these views not only by what I saw in the building of lasers and the exploitation of SQUIDS (supercon-ducting quantum interference devices) but also by my philosophy hero Otto Neurath. Neurath spearheaded the 1930s' unity of science movement of the Vienna Circle. But his idea of unity was not that physics – or anything else for that matter – could produce predictions by itself. He argued for unity *at the point of action*: We must bring the requisite sciences together as best we can, each time anew, to achieve the projects we set ourselves, from building a laser or a radar or even – as Neurath believed we had the intellectual resources for – to organizing and controlling the roller coaster of the economy. And though he urged us to talk the same language whenever possible, he never believed this language would stretch far or last long or capture much of what the separate users mean by its terms. Neurath advocated not a shared language but a 'universal jargon'.

Consider the MIT World War II radar project. Designing the radar took the united efforts of mathematicians, physicists, engineers and technicians, each themselves expert in one small domain with a language of its own, put together by the urgency of war and often against their will. It took a year for them to be able to communicate well enough to build a usable device – a year and the redesign of the physical environment. The building used to be arranged floor-by-floor according to prestige, with mathematicians at the top. The radar project mixed researchers from the different disciplines at long tables on each floor, tables that reflected in their very geometry the components of the radar to be built. Success was achieved not by constructing a single language nor by translation but by face-to-face contact that allowed enough interchange to make a go of it.

This is the space of interchange that Galison calls the 'trading zone', where two tribes stuck in linguistic isolation end up surviving through trade and 'commerce' of vital concepts. But recall: each group in this trade has to maintain its own language and its own understanding within itself, even if out of kilter in various ways. Otherwise it could not produce the detailed well-founded results needed for the project to succeed. The large MIT team that built the radar spoke Neurath's universal jargon or what we have seen Galison describe as a kind of pidgin, with each group maintaining different understandings of the terms they used in common.

Many other studies by different scholars studying different aspects of the practices of physics in different places see much the same thing. Consider just one more. This one looks at the Balkanization of the high temperature superconductivity community.[8] Low temperature superconductivity is taken to be well understood now. Our understanding of it is based on the original 1957 BCS theory, due to John Bardeen, Leon Cooper and John Schreiffer. BCS derived a formula for the temperature at which a superconducting material makes a transition to the superconducting state and from that an approximate upper limit to the transition temperature of around 30 K. In 1986 new types of materials, like cuprates, broke down that wall and showed unusually high transition temperatures (up to a recent superconductor which exhibits an extraordinary transition of 200 K). Ever since then the condensed matter physics community has been struggling with the problem of how superconductivity arises in some materials at far higher temperatures than that of conventional superconductors.

'Cooper pairs' play a central role in the BCS theory. These are pairs of electrons that, contrary to normal behaviour, attract each other rather than repel, due to their interaction with the crystal lattices that structure the superconducting material. The BCS theory does not serve well to model high temperature not only because the high transition temperatures are not predicted, but also because within that theory we do not have an account of how the correlated electron behaviour can arise. There are theories, yes. But not one of them is without serious problems. And the theoretical disputes are vicious.

Most scientists agree that the electrons still pair in the new materials, as they do in BCS theory, but disagree on how. A variety of theories are on offer and the community is rife with dissent and disagreement. To this date, even the most popular and promising candidates are unsatisfactory. Each, however, is hotly defended by its advocates as the single promising theory we should be pursuing, and each is vigorously attacked by the advocates of opponent theories.

How can each group defend its theory as best when none is satisfactory and there are a variety of alternatives? One clash is over what is most important in a good theory: top down or bottom up. Is it more important that the account mesh with high theory, with what are counted as well established laws and principles, or that it be as descriptively accurate as possible? Nobel-prize-winning condensed matter physicist Philip Anderson is clear where he stands: Get the fit with theory right and the experimental results will come round your way. You can even expect that some results that seemed to oppose your theory do not once all the details are understood, or that it turns out there was a flaw in the experiments to begin with. Bernd Matthias, who discovered and created hundreds of new superconducting materials, took just the opposite stand. He maintained that a model that lacks internal consistency and a first-principle derivation has a good chance of leading to the true mechanism for the phenomenon as long as it gets the facts right. From there it may be possible to build a more principled account, if we need or want one (though for purely practical purposes we may not). His National Academy of Science biography explains that he worked a great deal from intuition and that his 'enthusiasm for science was fueled by an unabashed joy in discovering something new, particularly when it did not depend at all on theoretical input.'

What do we learn about laws of nature from this dispute? Anderson himself is an opponent of vertical reduction, arguing for a kind of emergence of new principles at different levels of organization – which allows a special kind of autonomy to his own field of condensed matter physics. Nevertheless, it seems that if we follow Anderson, we should suppose that the 'well-established' principles of high theory that rule in condensed matter physics rule with an iron hand. Ensure your model behaves as these principles dictate and the facts your model predicts will be the right ones. But Matthias attributed his great successes at predicting and intervening into the world – not only by finding new superconductors but with ferroelectricity and ferromagnetism as well – to ignoring the full force of these principles, to paying minute attention to the details of the facts and to the use of a broad swathe of knowledge of different kinds, especially a deep understanding of the periodic table of elements. Matthias's successes it seems relied not on autocratic wide-ranging laws but rather on a motley assembly of knowledge and practice.

The battles between the warring camps also illustrate Galison's thesis about incommensurability. Even where experiment comes to centre stage in these disputes, it generally does not play the role we hope for in adjudicating among the alternative accounts, and for just the reasons that Galison has highlighted: The advocates of different theories do not share a common meaning for the same terms even in this single narrow domain. One very notable example is the 'kink', which is an observed unexpected spike in the dispersion curve during photoemission studies. Same word, but different groups construct the kink differently from the same body of data. Or take the *phase diagram*, a type of chart that shows conditions at which thermodynamically distinct phases occur. Often each camp builds and presents its own phase diagram, which contains only a selection of observed features, thus creating a vast series of almost incommensurable theorizations.

If one believes in the autocratic powers of physics despite her repeated failure to rule by herself in even the best of circumstances, there is of course a ready response one can give to this mass of evidence from studies of how physics works when it works successfully to produce novel predictions and technologies: Blame it on us, not on the world studied by physics. The world is totally ordered under the rule of physics law, one may insist. It is we weak intellects who have so far produced only an incomplete and inadequate approximation to what Queen Physics is really accomplishing.

My reply to this mirrors the famous Scottish Enlightenment philosopher David Hume in his reply to those who maintain that despite the appearance of disorder and evil all around us, the world is nevertheless really well governed by an all-powerful and beneficent God. If we knew the world were so ordered, we could use human failings and other excuses to accommodate the evidence of how the sciences work. But that is not the simplest, most natural conclusion to draw in the face of the evidence. If we must speculate about the structure of nature, I recommend we stick as close to the evidence as possible.

Whence order?

You will see specific accounts of how order emerges in special domains in physics, biology and in the study of society in Section 2 of this book. Here I take the liberty of sketching a brief picture I have come to see from my own studies in physics and economics of the kinds of order the world exhibits and what, if anything, is responsible for them. These studies suggest that we do not need to postulate just endless lists of *ceteris paribus* laws – though indeed, these are the only laws that have good evidence for their truth. Often we can trace these laws to a source and generally a source whose parts interact in intelligible ways, like the innards of the Strasbourg clock, to give rise to visible order.

In *The Aim and Structure of Physical Theory* the early twentieth-century physicist and historian of science Pierre Duhem distinguishes two kinds of minds: ample and deep. The ample mind 'analyzes an enormous number of concrete, diverse, complicated, particular facts, and summarizes what is common and essential to them in a law, that is, a general proposition tying together abstract notions.' The deep mind 'contemplates a whole group of laws; for this group it substitutes a very small number of extremely general judgements referring to some very abstract idea.' He continues, 'In every nation we find some men who have the ample type of mind, but there is one people in whom this ampleness of mind is endemic; namely, the English people.' The English can keep a thousand different concrete principles in mind at once, without getting muddled up. On Duhem's account their science has no need to replace these with a handful of very general very abstract principles – i.e. with the universal laws that physics is often thought to aim for.

I began my career, in the book titled *How the Laws of Physics Lie*, with reference to Duhem's national differences. God (or whatever deep principles underlie the order of nature), I maintained, has the mind of the English. If this is the way the English can do science, why should it not be the way nature does it herself? The natural world may be law-governed through and through but the laws that govern it look to be long and complex – at least if, as good empiricists should, we insist on laws that physics' successful predictions provide strong evidence for. These laws are many and diverse, hard to keep in mind and difficult to use – unless one has 'the mind of the English'.

Our best evidence still supports that God has the mind of the English. But today I add, 'Moreover, it's the mind of an English engineer!' I mentioned that Otto Neurath is a philosophical hero of mine. Here I take a cue from Neurath's scientific utopian vision of so-called 'social engineering', a dream implicit in his *International Encyclopedia of Unified Science* and many of his other various social and scientific projects. Neurath says, 'We could as social engineers also think of new types of patterns just as technical engineers discuss machines which do not exist up to the present.'[9]

So, God is an engineer – but not a mechanic, where I use the terms in the nineteenth-century sense. The nineteenth-century distinction is between *engines* and *mechanisms*. 'Engine' implies productive power, whereas a 'mechanism' is a device for executing a repetitive motion. Automata exhibit mechanical behaviour: repetitious and without creative power. Engineers learn about the powers things have – the elasticity of materials, that magnesium diboride can super-conduct at temperatures below 39 K, the power of a lever to raise large weights near the fulcrum by small weights far from the fulcrum. They use their knowledge of these powers and how their effects combine to construct devices that can give rise to totally new phenomena. Order, as I see physics and economics explaining it, where it exists, results from clever engineering, by nature, God, evolution, even unusually by happenstance or by us.

Engineered order is, however, a delicate thing. It is both local and fragile. *Locality* first. As the arguments I have been rehearsing suggest, little if any of the order we know about exhibits the regularities dictated by laws that apply universally. The generalizations that express visible order are generally only true locally. They are local to the underlying structure that gives rise to them and to which they are bound, an arrangement of things with powers that can work together to produce regular behaviour, which are often nowadays called 'mechanisms'. (I don't use that term because it has different meanings for different people. I call them instead 'nomological machines' because they give rise to predictable regular behaviour of the kind we learn about and record in our sciences, in engineering and in technology.) Consider for instance the myriad cause-effect relationships we rely on in day-to-day life. First thing in the morning I press a lever on my toaster and it delivers my breakfast. Later I may step on the lever on the floorboards of my car and it propels me out of my driveway. Manipulating the same cause does not produce the same effect given different gener-ating structures (in my language, different nomological machines).

Most visible regular order is also *fragile*. The generating machine breaks easily. Acorns have just the right internal structure to grow into oaks and they regularly do so – but not if they are smashed or overwatered or planted in the desert or snarfed up by pigs. Many are especially vulnerable to our attempts to exploit them. This is just what Chicago School economists like Robert Lucas claim about even well-confirmed principles involving macroeconomic quantities. Attempts to use these principles to intervene in the economy are likely to fail. Why? Because the interventions are likely to destroy those very regularities. How? By changing the underlying structure of behaviours giving rise to them. People see the interventions that the government is making and change their preferences and behaviours, which then no longer give rise to the original regularity.

Or consider John Stuart Mill in *On the Subjugation of Women* and in his disputes with Comte on the same subject. There may be a well-established regular association between being a woman and being inept at leadership, reasoning and imagination. Happily this regularity is not stable under changes in the nomological machine that generates it. Change the social structure so that the upbringing, education and experiences of women are like those of nineteenth-century British middle-class men and the regularities relating sex to leadership, reasoning and imagination will shift dramatically.

To avoid breakage, machines often come with shields, like the casing of a battery and the shell of an acorn. Happily it seems the planetary system carries on undisturbed without a shield but that is unusual. In any case the order generated depends on the proper arrangement and interaction of the basic features of the generating 'nomological machine' and can only be relied on so long as these arrangements remain stable – or have built in back-up arrangements in case of failure. Visible order then is very different from the stable and global order that would be exhibited by systems following universal and enduring laws of nature.

So the order that nature comes with is fragile. It can be broken – and we can break it: often for the worse, as in our influence on the climate; but often for the better – damns to prevent floods and droughts, antibiotics to stop the growth of germ colonies … Equally, when outcomes and arrangements are not predetermined by laws, we can create whole new systems of order, from toasters to dictatorships, again sometimes for the worse but often for the better.

The contingency of events that can be allowed once the universal rule of law is rejected leaves space for us to play an active role in what the world will come to be like. But that is not enough by itself. If what happens just happens, we have no more control than we do if all is fixed forever from the start. I urge that rather than laws, what science really learns about are the powers that things in nature have; order results from stable arrangements of features with different powers acting in consort. Steve Horst presents a similar view in his chapter 'From Laws to Powers' (Chapter 8). But I should note: to enable human beings to change the world and build new systems of order within it, it is not enough to endow us with powers. I have a mass and thus the power to attract other masses. But that will not help me to build a better – or a worse – world if I cannot exercise that power myself. We may indeed have a myriad of powers but if they exercise themselves willy-nilly when the occasion is ripe, we have no control. So we need in addition some kind of agency, as Tim Mawson argues in 'Freedom and the Causal Order' (Chapter 7): We are agents endowed with a variety of powers *and* with the ability to exercise them – to change arrangements in the world and to invent and set up new arrangements with new previously undreamt of outcomes. In a world where everything is not preset by law and where humans have genuine agency, it is not only God who is an engineer. We too are engineers. And we bear responsibility for the systems we create – and those we fail to create.

In conclusion

Let me turn for a few closing sentences to a preview of the issues in Parts 3 and 4 of this book, about the relations between mind and nature and between action, freedom and the rule of law. I have rehearsed a number of arguments that conclude that the idea of 'law' in the traditional sense of something eternal that governs everything in its domain entirely by itself with no exceptions allowed seems out of place in modern biology, even those domains of biology that provide impressive explanations and solid predictions. What does the job instead is local, piecemeal, contextual, developmental, holistic.

I then reviewed reasons to think that in these respects physics is not all that different. In particular physics is not enough, even in its own domain. Physics does not act on its own but rather in teams, co-operating with bits of knowledge at all levels, from a mix of sources with a hodge-podge of languages and concepts. So, the world of physics, when looked at through the lens of its most successful practices, is a dappled world. But if the world of physics is dappled and outcomes even in physics' own domain are governed by a motley assembly of features from different realms of knowledge, then there is causation from above, from below and from a thousand angles at the side.

Does this leave room for features of mind not determined by physical features? Does it leave room for these features of the mind to cause physical features? Does it leave room for free will? Or for Divine action in the world, without violation of the natural order? Those are questions we take up in Parts 3 and 4. What the arguments here teach is that the assumption of the autocracy and universal dominion of physics (or of any other discipline for that matter) should not play any role in settling it since this assumption itself does not sit well with much of contemporary scientific practice.

Notes

1   Mitchell 2002: 334.
2   Mitchell 2002: 333.
3   Mitchell 2002: 331.
4   Thanks due to Martin Carrier for this list.
5   Lange 2000: 162–3.
6   Earman, Roberts and Smith 2002: 284.
7   Galison, in Krige and Pestre 1997: 675.
8   Di Bucchianico 2014: 291.
9   'International Planning for Freedom' (1942) in Neurath 1973: 432.

References

Di Bucchianico, M. (2014), 'A Matter of Phronesis: Experiment and Virtue in Physics, A Case Study', in A. Fairweather (ed.), *Virtue Epistemology Naturalized* (Switzerland: Springer International Publishing).

Earman, J., Roberts, J., and Smith, S. (2002), 'Ceteris Paribus Lost', *Erkenntnis* 57: 281–301.

Galison, P. (1997), 'Material Culture, Theoretical Culture and Delocalization', in J. Krige and D. Pestre (eds), *Science in the Twentieth Century* (London: Routledge).

Lange, M. (2000), *Natural Laws in Scientific Practice* (Oxford: Oxford University Press).

Mitchell, S. (2002), 'Ceteris Paribus – an Inadequate Representation for Biological Contingency', *Erkenntnis* 57: 329–50.

Neurath, O., (1973), *Empiricism and Sociology*, R. S. Cohne and M. Neurath (eds.) (Dordrecht: Reidel).

Part Two

New Modes of Order in the Natural and Social Sciences

3

From Order to Chaos and Back Again

Robert C. Bishop and Roman Frigg

Editorial Link: Roman Frigg and Robert Bishop – trained in both physics and philosophy – take examples from physics to show how order is emergent, holistic and contextual, not produced by universal and exceptionless laws. For example, in chaos theory there is a sensitive dependence on initial conditions, which themselves cannot be wholly specified with complete precision. Often 'the system establishes the pattern', rather than the system being built up just from the independently calculated movement of its parts. So there is self-organization and patterns of emergent order in the universe, which the old model of invariant laws and precise initial states cannot explain. KW

Whether it is the movement of heavenly bodies, or the arrangement of pots and pans, or global politics or family life, we characterize almost everything as either ordered or disordered. What do we mean when we say that there is order in a domain? A venerable tradition in physics associates order with the presence of all-encompassing exceptionless laws. Ever since Galileo set out to describe the motion of falling bodies in terms of simple mathematical expressions, physicists aimed to discover laws of ever greater scope and generality. A domain is then regarded as ordered if its objects are seen as behaving according to a general law. In this essay we articulate this conception of order, describe how it crumbled in the wake of chaos and complexity studies, and portray the emergence of a new conception of order focused on patterns of self-organization.

Newtonian order

This law-based conception of order reached its first peak with Newton's mechanics, which he formulated in his monumental masterpiece *Philosophiae Naturalis Principia Mathematica* (*The Mathematical Principles of Natural Philosophy*) published in 1687. At the heart of Newton's theory lies what is now known as *Newton's Law of Motion* ('Newton's Law', for short). The equation describes how a force changes an object's state of motion (e.g. wind lifting a kite skywards). An object's state of motion is its position and its velocity: if you know where an object is and how fast it moves in a particular direction, then you know all there is to know as far as mechanics is concerned. Changing a system's state of motion amounts to changing its position and velocity. The rate of change of the velocity is known as acceleration. When you get into your parked car and start driving you accelerate: you change its velocity from zero to 30 miles an hour within the space of five seconds.

The great insight behind Newton's equation is that every change in velocity is the result of forces acting on the object. More specifically, the equation postulates that the acceleration of a body times its mass is equal to the total force acting on the body. Expressing this insight using mathematical symbols we get $F = ma$, where F is the total force, m is the mass of the body and a is the acceleration of the body. The equation allows you to calculate, for instance, the acceleration of your car as a function of the power of the engine, and then it tells you where your car will be at any later time. And this is by no means limited to cars. Newton's equation is in effect a recipe for determining the position and the velocity of any physical object at some later time t given its current state of motion and the forces acting on it.

While this can be made mathematically precise, Newton's equation works like this: Suppose there is a car stopped at noon at an intersection where a very long, straight stretch of highway begins. We want to know where the car will be 15 minutes later. The Newtonian recipe is to determine the total force F acting on the car (the power of the engine and air resistance are the main factors), determine the car's mass m and then plug these into Newton's equation. The solution of this equation tells us the position and velocity of the car 15 minutes later or at any time t we want to know. The original position of the car at noon is called the car's *initial condition*.

At first sight, Newton's Law may appear to be a source of bewilderment rather than order. To see how a conception of order emerges from it we first have to look at how it is put to use. From the point of view of Newtonian physics, the material world is just a huge collection of material particles, which are subject to forces. Everything that there is – chairs, rivers, clouds, tornadoes, galaxies – is treated like a swarm of particles, where change in the world is change in the fundamental particles' state of motion, just like the change in state of motion of our car. Clouds change their form because the particles of which the clouds are made up change their positions and velocities, tornadoes move because the particles of which they are made up move, and so on. In such a universe, Newton's Law is truly general: it applies to absolutely everything. The syllogism is as simple as it is striking: Newton's Law governs the motion of particles subjected to forces; the universe is entirely made up of particles which exert forces on each other; therefore Newton's Law governs the entire universe. So $F = ma$ is the equation of the universe.

But merely subsuming everything that happens under $F = ma$ seems to be at best the beginning of an explanation for order in the physical world. Why should we think that there is order in the way waves break on rocks, the way leaves drift around in stormy weather and the way galaxies move in space just because they are instances of $F = ma$? Merely being subject to an exceptionless and general law is not enough to qualify as being ordered. The proverbial 'que sera sera' is a general and exceptionless rule, yet no-one would take it to be a source of order.

So there must be something to Newton's Law in addition to generality and exceptionlessness that makes it a source of order. Determinism seems to be a good candidate to be the hitherto missing extra ingredient. A law is deterministic if it uniquely determines the evolution of a system: given a system's state of motion – the positions and velocities of all the particles that make up the system – the system can evolve only in one way. So if Newton's Law is deterministic, then the state of a system at *any* other time is uniquely determined by the system's current state.

To make this more vivid, consider the following thought experiment. Suppose you have a thousand copies of the same system. Pick a particular instant of time and make sure that at that time all thousand systems are in the same state. Now you let each system evolve according to Newton's Law and then look at the systems again one hour later. You will find that they all have evolved in exactly the same way to the same state! This will be so with any number of copies; you can have 100,000 or 1,000,000 copies of the system and you will always find that all of them evolve in exactly the same way. Newton's Law governs the world as a whole and the consequence of determinism is that things can go in only one way. The course of the world is written in Newton's equation and the initial state and all the world can do is follow its predestined path – the world is a clockwork ticking away without ever having the possibility of changing its course.

Determinism is a crucial element of the Newtonian conception of order, but there is one last bit missing. Determinism eliminates randomness and chance from the picture – things happen in a particular way and no deviation from the prescribed trajectory can ever occur. This certainly does have a ring of order to it, but it's not enough yet because determinism is compatible with very complicated and unwieldy kinds of motion. A deterministic system can jump around wildly and uncontrollably, which seems incompatible with our sense of order.

The last stone in the mosaic of Newtonian order was put into place by French physicist and mathematician Pierre-Simon Laplace in 1814 with a famous thought experiment. The protagonist of this thought experiment is a supreme intelligence which is now known as Laplace's Demon. The demon is a remarkably skilful super-scientist. It is able to identify all basic components of the world and the forces acting between them, and observe these components' positions and velocities. It formulates the Newtonian equations of motion of the entire world, and then uses its infinite computational powers to solve these equations correctly in no time. If everything in the Newtonian cosmos is reducible to the motion of particles, knowing the solutions to the equation of the world amounts to knowing literally everything that there is to know about the world. This led Laplace to conclude that the Demon is in the enviable position of omniscience because nothing would be uncertain to it and the future as well as the past would be present instantaneously to its eyes.

The crucial point in Laplace's thought experiment is that determinism is equated with predictability: because the equations have a unique solution, solving the equations is tantamount to generating a prediction. If determinism and predictability are the two sides of the same coin, then one can know the future behaviour of any deterministic system with certainty.

Hence, the Newtonian notion of order is one of a world being governed by a universal law that guarantees complete predictability, thereby removing any uncertainty from the picture. Dealing with an ordered system means dealing with a system about which one knows what's going to happen. This idea of order has been influential beyond the confines of mechanics, playing a crucial role in other parts of physics. For example, thermodynamics and statistical physics are cases in point. At the heart of these disciplines lies the concept of entropy, which is understood as a measure of order: the lower a system's entropy, the more ordered it is. A careful look at the definition and uses of entropy reveals that it quantifies how much information about a system we are missing: the higher a system's entropy the less we know about it and the more uncertain we are about its true state. So, again, we have a notion of order that equates order with the absence of uncertainty.

The Newtonian picture has a further benefit that adds to the perception of the Newtonian universe as one that is ordered. Early applications of the theory, most notably in astronomy, provided beautiful mathematics. Using Newton's equation to calculate the motion of planets shows that planets move in elliptical orbits, and the equation of an ellipse can be written down in a single line on a piece of paper. In the same vein, when subjecting other mechanical problems to a Newtonian treatment physicists found that the sought-after solutions were mathematically beautiful and so a little cousin was added to the notion of order as predictability: order as motion that is mathematically beautiful.

Newtonian chaos

What looks too good to be true usually is too good to be true, and the Newtonian picture of order is no exception to this rule of common sense. We are now going to see how the Newtonian notion of order crumbles. To see this, let's recapitulate Laplace's thought experiment by listing the Demon's essential capabilities:

1   It can specify the *true* initial conditions of the world.
2   It can instantaneously calculate the *true* solution of every equation.
3   It can formulate the *true* equation of the world.

A look at the history and philosophy of physics reveals that each of these conditions is problematic as soon as it is taken out of the context of a thought experiment. Consider the first condition. Not being creatures with magic microscope eyes, we cannot simply 'read off' the precise location and velocity of a particle by looking at it; we have to perform measurements. However, measurements always have a finite precision. Think of the thermometer in your living room. When the thermometer shows 67 degrees Fahrenheit, you would make a mistake if you took this to indicate that the temperature in the room is *exactly* 67 degrees Fahrenheit. Typical thermometers have a precision of about ±1 degrees Fahrenheit, which means that the temperature in the room is somewhere between 66 and 68 degrees Fahrenheit. Good instruments are more precise than bad ones, and investing in a high precision laboratory thermometer will bring down the measurement uncertainty considerably, but there always are limits to precision. There are no thermometers that give us *exact* values. What is true of thermometers is true of measurement devices for position and momentum. The best we can ever get is the information that the value we are looking for lies within a specified range. That range can be very small, but it will never be zero.

Why should we find this troubling? Surely the results of high precision instruments are good enough for our purposes! The thinking behind this is that changing the initial conditions a little bit will not affect the outcome very much. Returning to our car example, from an initial starting point *A* the car travels to a destination point *B* via a particular path. With the initial conditions and the force acting on the car, we can use Newton's equation to calculate the car's path – its trajectory. Now suppose the car starts from a slightly different starting point *C*. Given the same force, Newton's equation would give us a new trajectory and tell us that the car reached a destination point *D*. If the initial starting points *A* and *C* are very close to each other, then the destination points *B* and *D* would be very close to each other. And the two trajectories also would be very close to each other for all times. In other words: similar initial conditions lead to similar trajectories. This is known as the *strong principle of causality* ('strong principle' for short).

This principle is intuitive and it holds true in many everyday contexts – if you let a ball drop one inch to the left, then it lands one inch to the left. For this reason the strong principle has been accepted as a truism for centuries and measurement imprecision has been considered as being of no theoretical interest. To everybody's great surprise, this has turned out to be a fatal mistake. French mathematician and physicist Henri Poincaré and American meteorologist Edward Lorenz discovered (independently of each other) that small changes in the initial condition can have huge effects if a system's equation has the mathematical property of being *non-linear*. The exact meaning of 'non-linear' need not occupy us here; the important point is that most real forces are non-linear (for instance, gravitation is non-linear) so non-linearity is the norm rather than an exception that is easily dismissed.

The consequence of this is that in most systems the strong principle does not hold: even if the initial conditions are very close the trajectories can differ dramatically from each other. This fact is known as *sensitive dependence on initial conditions* ('sensitive dependence', for short) or *chaos*; in popular parlance it is sometimes referred to as the *butterfly effect*: a small change in the initial conditions (a butterfly in China flapping its wings) can cause a huge effect (a tornado in the Caribbean). So we cannot infer from the fact that initial conditions are similar that the later trajectories will be similar too. In general, we can only uphold the *weak principle of causality*: exactly identical initial conditions lead to the same trajectories, which is of course a straightforward consequence of determinism.

The failure of the strong principle has wide ranging consequences. These consequences are explored in the fields of *chaos* and *complexity studies*, and we will examine some of these insights later. For now let's continue our discussion of order. The disconcerting consequence of sensitive dependence on initial conditions is that we can no longer dismiss measurement errors as irrelevant. If the slightest difference between initial conditions can make a huge difference at later times, then measurement errors do matter. This has dramatic consequences. In a system with sensitive dependence obtaining exact initial conditions is necessary to predict the future accurately because if the conditions we use to make our calculations are only slightly off, then our calculations lead to the wrong result. But taking measurement inaccuracies seriously, as we should, we can never obtain an exact initial condition. Hence we cannot predict the future with certainty. This pulls the rug from underneath the view we have been discussing. We associated order with determinism because, following Laplace, we associated determinism with predictability. But chaos now tells us that this association was faulty. An important aspect of our notion of order turns out to be false.

Chaos also undercuts the second condition. There were many early successes in astronomy and other simple mechanical systems such as pendula and projectiles. Such successes, practical problems with solving equations notwithstanding, led to the view that solutions of Newton's equation could eventually be found and that they would be mathematically nice in something like the way in which the ellipses of planetary motion are mathematically nice. This expectation is frustrated by chaos and complexity, which also furnish us with the insight that even though solutions exist, they are not of the kind that one can write down with pencil and paper as a combination of a finite number of known functions. This means that solutions to non-linear equations are not available to us in a manageable mathematical form. The only way to get our hands on (an approximation of) solutions is to solve equations numerically on a computer – but even where computer solutions are available (more complicated equations are beyond the reach of even the most powerful computers), they are less informative and less useful than explicit mathematical solutions. So chaos not only prevents us from being able to make relevant predictions; it also renders solutions in useable mathematical form unattainable.

The third condition is also beset with problems, although the problems are of a somewhat different nature. The first two conditions were undermined from within, as it were, because it turned out that for purely mathematical reasons one could not do things we initially thought we could do. The third condition fails for extra-mathematical reasons. The notion of writing down the equation of the entire world bears all the imprints of a thought experiment. It is one thing to imagine a demon able to write down the Newtonian equation for the entire world; it is a different thing to actually do it. Successful applications of Newtonian mechanics deal with relatively small and well-circumscribed systems such as the paths of projectiles and the motion of planets. However, we are very far away from being able to formulate true equations for relatively larger systems, and the world as a whole does not even seem to appear on the horizon of workability.

The hardnosed reaction to all these points would be to dismiss them as merely practical limitations. That we cannot measure initial conditions exactly, nor handle complicated mathematical solutions, nor write down equations for complex and large systems, has only to do with our own limitations as human beings. But human limits are irrelevant for the order of the world in itself: there are true initial conditions, there are solutions to complex equations and there may even be a true equation of the world – that we are unable to get our hands on it is neither here nor there as far as cosmic order is concerned. The world is what it is irrespective of what we know about it and irrespective of how well we can describe it. The above arguments therefore only show that we are never able to fully grasp the order of the world; they don't undermine our concept of order.

This response faces two difficulties. The first is that it is at least controversial whether the limitations are only practical. Mathematician and physicist Max Born defended the view that the very idea of a precise initial condition is a mathematical fiction without physical meaning; we are therefore chasing rainbows when looking for exact initial conditions. Poincaré thought it an impossibility to ever get a handle on solutions of complex non-linear equations with paper-and-pencil mathematics simply because that mathematics is the wrong tool to understand non-linear systems. He, therefore, advocated a shift in focus to qualitative features of systems' dynamics (more about that below). Several philosophers, for instance Nancy Cartwright, have argued that seemingly universal and exceptionless laws are in fact not so. First appearances notwithstanding, the domain of applicability of laws such as Newton's is limited to relatively simple mechanical systems, and we have no reason for believing that these laws apply outside their initial domain of successful application.

Even if these disputes could be resolved in favour of those who see said failings as practical limitations of no fundamental import, we would still be left with another problem. The starting point for an investigation into order is the basic fact that at least large parts of our environment seem to be *ordered to us*. Important parts of the world of everyday experience, as well as systems studied in the sciences, not only *appear* ordered; they have an order we can understand and exploit to make successful inductive generalizations. It remains a mystery why this is so if we coin a notion of order that is relevant only to demons and remains forever concealed from mortals who are saddled with too many cognitive limitations. Even if such a notion of order can be maintained as a philosophical concept, it is useless in explaining why we experience the world as ordered and why we can operate successfully in it. Indeed, our everyday experience in the world as well as our scientific investigations seem to undermine the validity of such a metaphysical conception of order.

Self-organization

So we need a different notion of order, and fortunately the study of non-linear systems suggests a different way of thinking about order. At its most basic, order is a relatively stable and identifiable macroscopic pattern of organization. Such a pattern can be spontaneous, self-generated, or externally imposed. Engines and watches would be examples of externally imposed order. Their parts are pieced together into a specific order by humans or machines programmed to create or impose a specific kind of order on the component parts.

In contrast, *self-organization* or *self-organizing behaviour* is a form of order that arises spontaneously or is self-generated. At first this may seem counterintuitive since no intentional co-ordination is involved and often our expectation is that disorder arises in the absence of such co-ordination. Yet, the formation of planetary systems, galaxies, tornadoes and hurricanes are examples where matter self-organizes into coherent patterns in the absence of intentional co-ordination. And the most surprising thing is that many patterns of self-organization turn out to be the result of equations like Newton's!

We can illustrate these ideas with the example of the so-called Rayleigh-Bénard convection, which is the physical mechanism underlying familiar phenomena such as the formation of cloud streets in the sky and the hexagonal patterns in volcanic basalt. A typical arrangement for Rayleigh-Bénard convection is a fluid such as water in a chamber with a temperature difference between the top and the bottom of the chamber.

The bottom is heated so that it is maintained at a higher temperature than the top. This creates a temperature difference, where the fluid near the bottom is hotter than the fluid near the top. As long as the temperature difference between the top and the bottom remains small enough, the extra heat in the warmer fluid tends to

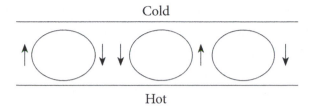

Figure 3.1. The formation of convection cells. From Chapter 3: Physics – From Order to Chaos and Back Again.

spread evenly and the fluid seeks to come to a uniform temperature throughout the system. Hence, as long as this extra heat can spread more rapidly than the warmer fluid can rise, the system remains in a stable state such that the fluid near the bottom remains at the bottom.

The situation becomes very interesting as the temperature difference between top and bottom becomes larger. Just like warm air rises while cool air sinks, the warmer fluid at the bottom of the container tries to rise upward while the cooler fluid tries to sink downward. As more heat is applied to the bottom, the fluid at the bottom of the chamber becomes hotter faster than the heat can spread and so the hotter fluid's tendency to rise upward steadily increases. When the temperature difference between the bottom and top becomes large enough, a critical point is reached, which marks the onset of large-scale pattern formation.

Hot liquid starts moving upwards until it comes close to the (cooler) top where it cools down. Cooler liquid has a tendency to move downwards. It is pushed sideward by the stream of ascending hot liquid and the result is a vertical circular motion, which is known as a *convection cell*. This pattern repeats itself throughout the liquid, which is now structured as a string of convection cells. This is illustrated in Figure 3.1. Depending on the details of the system, very different patterns can emerge. Wind and ocean currents, tornadoes and hurricanes are just a few examples of convection structures, where different patterns are the result of different physical conditions. All of these convective structures involve temperature differences similar to the model system in Figure 3.1.

This is an example of self-organization. The liquid isn't forced to form this pattern by any external agent; the pattern evolves spontaneously and without aid from the outside. The system as a whole establishes the pattern of convection cells. The precise shape of the cells depends on the details of the set-up such as the shape of the walls forming the chamber and the viscosity of the fluid. In principle the fluid could flow in an arbitrary number of directions (imagine suddenly removing the chamber walls and the fluid flowing in all directions onto the laboratory floor). Depending on the details, the convection cells may form cylinders, squares, hexagons or other shapes.

Although just one example, Rayleigh-Bénard convection illustrates a number of general features of self-organizing behaviours found in nature:

- The pattern of the convection cells is due to a combination of external constraints (size and shape of the container) as well as the dynamic behaviour of the fluid itself.
- The newly organized order state has a discernible nesting of organized structures: there are discernible macro patterns such as convection cells and at the same time the motion of the molecules of which the liquid consists must follow particular paths.
- The interrelationships among these different structures involve intricate couplings between the larger-scale structures and those on smaller scales. In our example, the convection cells determine which individual fluid molecules will remain within any particular convection cell or migrate to another cell.
- The macro structures are resilient under small changes. In our example, the convection cells are stable under small changes in

the temperature difference between the bottom and top, and are adaptive under moderate changes in temperature difference. For instance, if the heat applied to the bottom is raised a moderate amount, the cells may begin to oscillate in complex ways or even change shape (e.g. from a hexagon to a cylinder).

A new kind of Newtonian order

An important point in all this is that we have not simply thrown Newtonian physics overboard and made a fresh start. Quite the opposite. When we study the structure of convection cells theoretically we describe the liquid with Newtonian equations (or to be more precise: the so-called Navier-Stokes equations, which are the 'liquid version' of Newton's equations). As above, the equations are non-linear and show all the unpredictable behaviour that led to the downfall of the Laplacean paradigm. The key difference is that we are now using these equations in a very different way from how they were used by Laplace's Demon. Under the Laplacean paradigm, the emphasis was on the study of individual trajectories of the particles composing the system and order was associated with control over where these particles are at a given instant of time. The study of self-organizing behaviour focuses on large-scale patterns such as convection cells, and aims to understand the conditions enabling large-scale patterns to form and the nature of those patterns. Even though we have a set of equations describing the fluid and the constraints on it, we can only make a number of broad qualitative predictions, but no precise forecasts about, for instance, when the transition from the quiescent state to the convection cells will take place, or what particular shape the convection cells will have.

This is the shift from a quantitative analysis to a qualitative study of dynamic properties. When Poincaré announced this change in perspective at the turn of the last century, this was nothing short of a paradigm shift. The new paradigm has since become a widely practiced discipline. Much of the study of chaotic and complex systems involves qualitative understanding of the formation and nature of patterns such as Rayleigh-Bénard convection, and self-organization has become of wide interest to scientists.

For this reason, chaos and complexity studies do not provide many predictions of quantitative details; rather, they help us understand and predict qualitative features and patterns. Attempting to study individual trajectories of the molecules composing the fluid in Rayleigh-Bénard convection does not lead to any detailed predictions for the conditions under which such patterns form nor for the detailed shape of such patterns. In particular, focusing solely on the individual trajectories of molecules and their interactions with their nearest neighbours leaves out the effects larger-scale structures such as convection cells have on the individual molecules. The patterns and behaviours of convection cells are not fully determined by the properties of the individual molecules composing the cells. At the same time, the cells would not exist in the absence of the molecules. This kind of complicated relationship between large-scale structures and constituents composing those structures is characteristic of chaos and complexity studies. Although these kinds of relationships put many of the detailed quantitative predictions out of reach, we can still make a wide variety of useful predictions about the qualitative features of these systems.

Another general feature of chaos and complexity exhibited by Rayleigh-Bénard convection is that the order that self-organizes is mainly influenced by the local or particular context of the system. Concrete factors such as the size and shape of the container, the particular temperature difference between the top and the bottom, the particular kind of fluid in the chamber and the particular dynamic behaviour that arises in the form of large-scale convection cells are at least as important as any laws that may be coming to expression in the particular context. For example, the Newtonian laws involved in the interactions and motions of the fluid molecules alone do not fully determine the behaviour of those molecules. The constraints represented by the actual experimental arrangement (e.g. chamber size and shape) as well as those constraints associated with the large-scale convection patterns strongly shape the allowable behaviours of the molecules.

Whether the laws involved are universal or are local and particular is of no consequence to the behaviour of chaotic and complex systems. Whatever laws play a role in these non-linear systems, they are always qualified and shaped by the concrete contexts of the systems and their environments. So while there may be some universal laws at work in such systems, the effects of these laws are but some of the players in the arena established by the context of the systems and their history of development. This channelling of the laws by particular contexts is independent of whether these laws are expressed in equations derived from an overarching theory such as Newtonian mechanics or statistical mechanics ('top down') or derived by considering the local, particular situation ('bottom up'). The order present in self-organizing systems is the result of many more effects than just the outworking of a set of laws.

Physicists studying self-organizing behaviour therefore do not attempt to understand the order exhibited by non-linear systems solely in terms of universal, exceptionless laws. Even where such laws are available, they are only one ingredient of an explanation. Rather, they tend to focus on the particular contexts of the systems in question and seek to understand the systems' behaviours in those contexts. These scientific practices fly in the face of a traditional conception of the sciences as seeking to unlock the order of the universe through universal and exceptionless laws.

Self-organizing systems provide a model for how to understand order as arising from a complex set of interactions and contexts.

The laws and elementary particles of physics may provide necessary conditions for the order we experience. After all, if water molecules did not follow some mechanical laws of motion then convection cells would not emerge. But what we see in self-organizing phenomena such as Rayleigh-Bénard convection is that the elementary particles and the laws involved in their interactions are not sufficient to determine all the behaviours of fluid molecules. Only the combination of the concrete context in which the convection cells arise with whatever laws are involved with the molecules themselves leads to the necessary and sufficient conditions for the actual behaviour of fluid molecules in such convection systems.

For instance, without gravity there would be no tendency for warm fluid to rise and cooler fluid to sink. Yet, gravity by itself does not determine the actual rising and sinking of fluids. That requires additional conditions such as the densities of the particular fluids, the sources of heat for temperature differences, constraints such as chamber walls or the surface of the Earth and a multilayered upper atmosphere, among others. So while what we consider to be the most fundamental laws and particles of physics contribute necessary conditions for there to be phenomena such as Rayleigh-Bénard convection, particular concrete contexts contribute the rest of the conditions sufficient to produce the ordered convection patterns we observe. There are no violations of the laws involved in the interactions of elementary particles and molecules since these laws only contribute some of the conditions necessary for the behaviours of particles and molecules. The laws, particles and contexts work jointly to produce the order we observe in the universe.

Such a picture of emergent order generalizes nicely. The particular beginning our universe had involved a set of laws, matter and a context where the combined conditions promoted the formation of galaxies. This context contributed conditions enabling the self-organization of the Milky Way galaxy in particular. The context of the Milky Way galaxy provided conditions promoting the formation of stars and planets. A particular local region of the galaxy had conditions promoting the formation of the Sun and our solar system. In turn, the formation of our solar system had appropriate conditions promoting the formation of a planet with a liquid core in an orbit in the habitable zone around the Sun. This context provided the conditions promoting the formation of complex biomolecules and, eventually, life.

Such self-organized order does not depend on the laws involved being universal and exceptionless; nor does it depend on determinism, absolute predictability or any other features of the old Newtonian, law-governed view of order. Instead, this new picture of order depends on the particular context of our universe and the localized contexts in particular regions of the universe.

Further reading

Bishop, R. C. (2011), 'Fluid Convection, Constraint and Causation', *Interface Focus* 2012: 4–12.

Kellert, S. H. (1993), *In the Wake of Chaos* (Chicago: Chicago University Press).

Ruelle, D. (1991), *Chance and Chaos* (London: Penguin Books).

Smith, P. (1998), *Explaining Chaos* (Cambridge: Cambridge University Press).

4

Multiple Orders in Biology

Eric Martin and William Bechtel

Editorial Link: Bill Bechtel and Eric Martin, who study biology in practice, point out that biologists do not appeal to universal laws of nature in the same way that physicists have tended to do. Systems of classification in biology are contested, and this suggests that they do not simply copy 'nature's joints', but reflect contingent and largely pragmatic decisions and aims of the biologists. Much biological work is holistic, requiring both reductionist (old-style) analysis and holistic reconstruction (in terms of the whole systems within which parts have a function). This recognizes that the operation of individual parts is affected by their place in complex networks. The evolutionary principle of descent with modification is also complicated by the horizontal transfer of genetic material in specific contexts, suggesting piecemeal, local, historically dependent development in particular locations. All these factors point to principles of self-organization in local contexts as more important than the application of supposed universal 'bottom-up' laws as useful scientific models – though both continue to have their place in biology. KW

Ways of finding order in biology

On the face of it, the world of living organisms exhibits some striking differences from the non-living realm. Living organisms stand in a non-equilibrium relation with their environment and must extract energy and matter from their environments to build and repair themselves. Growth and reproduction are further hallmarks of most forms of life. Moreover, organisms regularly extract information about their internal constitution and external environment and alter their internal processes, external activities and even their genetic constitution in response. While these and a few other generalities apply across living forms, the specifics about how living organisms accomplish these things are highly varied. There are an enormous number of life forms, employing a myriad of mechanisms to maintain themselves and leave descendants. Overall, the biological world appears far from tidy but rather highly disordered.

Such variability in the natural world would seem to be an impediment to modern science, which has tended to emphasize the discovery of general laws. Following the successes of Newton, who advanced a few general laws that explained a wide range of phenomena, many philosophers saw physics as the model of scientific explanation to which all other inquiries should conform. Yet biology has been stubbornly recalcitrant to such approaches, leaving many to think that biological phenomena are of a qualitatively different sort from those studied in physics. Immanuel Kant famously asserted that there could never be a Newton of a blade of grass. His scepticism for the prospects of a scientific biology arose from the fact that organisms display characteristics that machines never could, such as interdependence and self-generation.

Kant's favoured solution was to recognize teleology – an irreducible goal-directedness or purposefulness – in all biological systems. Teleological explanations have enjoyed only marginal scientific status following the introduction of evolutionary theory, but the question of biological order in the scope of the natural world has continued to exercise many thinkers. Biologists have developed several strategies for finding order amongst living organisms, of which we will focus on three – taxonomies, mechanistic explanations and evolutionary patterns of descent with modification. We will argue that these strategies have provided important organizing principles for biologists. We will also show, however, that even as these tools bring some order to biology they also give rise to complexities that frustrate any hope of some manageably simple mode of order. Thus, biology manifests multiple taxonomic schemes that serve different scientific objectives, mechanisms organized in ways that generate complex behaviour and descent relations that involve horizontal as well as vertical exchanges of genetic material. While seeking order, biologists have to adapt their aspirations to the phenomena they seek to understand.

Sorting things out

One way biologists generate order in nature is by developing classification systems. By placing things in categories, we acquire ways to name and make generalizations about them. We can make certain inferences based on knowing that some creature is a vertebrate, and even more specific inferences if we know its exact species. Most biological classification is hierarchical, sorting groups and sub-groups into ever smaller and more specific categories.

Classification might seem like an uncontentious activity, something akin to stamp collecting, which has scientists unanimously agreeing to place particular objects within their most natural groupings. A close examination of scientific practice, though, reveals that such groupings are neither undisputed nor particularly natural. This is true not only in new biosciences, but even in the most traditional areas of science, like astronomy. After being classified as a planet since its discovery in 1930, astronomers in 2006 demoted Pluto to the new category of dwarf-planet, because of updated ideas about what it means to be a planet. (One of the scientists responsible for the reorganization wrote a book called *How I Killed Pluto*.) Biologists have dealt with such taxonomic debates for a very long time.

The eighteenth-century Swedish naturalist Linnaeus inherited from the Ancient Greeks the belief that behind the variable appearances of organisms are essences in virtue of which an organism is what it is. Linnaeus developed a classification scheme that arranged organisms into kingdoms, orders, families, genera and species based on objective similarities in these essences. The essence shared by members of a species provided necessary and sufficient conditions for an individual to be a particular species. The thought here, as with other putative natural kinds such as the periodic elements, is that some particular essence – a unique trait – will be found that can identify an organism as a member of a kind, providing an objective ordering principle for the classification.

Species remain the foundational unit of classification but, following Darwin, they are also the unit of evolution: it is species that evolve, not individuals or higher taxa. But this second role for species has posed a challenge to the essentialism underlying Linnaean classification since crucial to evolution is the idea that species can evolve into other species and that in the process any traits they possess can vary. On the most common criterion, once organisms differ from one another sufficiently that interbreeding is not possible then a species has split into two. Given the range of traits on which differences can prevent interbreeding, this perspective undercuts the assumption that there are necessary and sufficient traits defining species. It also explains why it has been so notoriously difficult to identify such traits and why the boundaries between species are often vague. The challenge to find order is further complicated by the fact that biologists appeal to multiple species concepts, including not just versions that focus on reproductive compatibility, but ones which emphasize shared evolutionary lineage and shared use of an ecological niche.

These different ways of organizing open the door for disputes between biologists about how order is best achieved. An extremely heated controversy emerged among systematists in the mid-twentieth century about how to organize categories above the species level. Influenced in part by positivism in philosophy, taxonomists known as *pheneticists* sought objective measures of similarity between species, especially with respect to morphological traits, while avoiding any speculative or theoretical considerations such as evolutionary descent. They produced dendrograms (see Fig. 4.1) designed to represent overall similarity: the shorter the lines from the branch point between two species, the more similar they are. These pheneticists were challenged by cladists (also known as *phylogenetic taxonomists*) who made descent relations primary in classifying organisms. They produced cladograms (Fig. 4.1, lower part) in which the branching structure represents speciation events. Species fall within taxa that group together ancestral species and all their descendants, irrespective of their similarity. While cladists were generally viewed as the victors in this controversy, the field remains highly contentious as variants of each approach are developed. While taxonomies offer ways to order species, there is far from a consensus as to what this order will be like.

Classification is not just controversial in the case of species. We briefly discuss two other examples where biologists have tried to establish order – among proteins and psychiatric disorders. Proteins are extremely important molecules in living systems since they enable most chemical reactions. Initially they could only be identified in terms of the reactions they enabled, but by the middle of the twentieth century investigators established that proteins were composed of amino acids bound sequentially into polypeptide chains, and began to identify the distinctive sequences of different proteins. The *Atlas of Protein Sequence and Structure* advanced a categorization of proteins into families, with families grouped into superfamilies and divided into subfamilies. As with species, there are multiple ways of organizing proteins. This is well illustrated by research on the nuclear hormone superfamily. This group of 48 proteins in humans

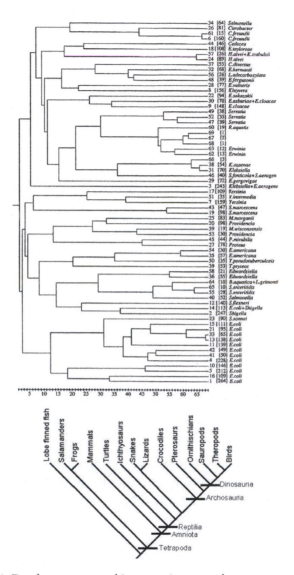

Figure 4.1. Dendrogram as used in numeric approaches to taxonomy and cladogram as used in cladistics approaches. Note that taxon D, which is shown as more similar to E in the dendrogram, is shown as sharing a branch point with C in the cladogram. Such differences are common between phreneticist and cladist taxonomies. Figure from Chapter 4: Multiple Orders in Biology, *Genentech's Access Excellence*, Eric Martin and William Bechtel.

is divided into families according to several different schemes. One scheme focuses on structural criteria, location of activity and mode of function. Another scheme focuses on the functions of the cells in which proteins are expressed, while a third emphasizes evolutionary homology. In this case there is little controversy about which system to use, but the fact that biologists appeal to whichever system is most useful for a given purpose reveals the complexity of the order that has been established.

Medical practitioners have long found it valuable to categorize pathologies so as to facilitate diagnosis, conduct research and organize treatments. In psychiatry this has generated considerable controversy, especially when the widely used categorization system, the *Diagnostic and Statistical Manual of Mental Disorders* (DSM), is undergoing revision. The reason for this is in large part that the categories are even more clearly conventional than those employed for organisms or proteins and there are few clear principles for including or excluding categories within the system. For example, the first two major editions of DSM, in use from 1952 to 1974, included homosexuality as a bona fide pathology. It was excised in DSM III in the wake of a widespread antipsychiatry movement in the late 1960s and attendant protests from the gay rights community. Classifications are also frequently added, including ones that expand the domain of the pathological to include common behavioural variations like shyness and grieving, whose construal as pathologies has sometimes coincided with the approval of new psychopharmaceuticals ostensibly treating those conditions. In the preparation of the most recent version, DSM V, compulsive shopping and internet addiction were proposed as new categories of pathologies. Deleted, though, was the separate classification of Asperger's Syndrome and other Pervasive Developmental Disorders (PDDs) in favour of a single category of Autism. It is difficult to see how the deliberations over the DSM could be construed as identifying 'nature's joints' (to use Plato's metaphor for finding nature's true divisions), when they appear to depend so strongly on acceptance of particular, contingent ways of living or decisions about what behaviours should be treated.

It should be clear at this point that classification systems do provide order but that scientists classify for different reasons – depending on their purposes, methods and objects of investigation – and consequently there are multiple different categorization systems in use. Taxonomies are not just bean-counting, nor are they mere semantics, stipulating linguistic usage. On the contrary, taxonomies and the categories on which they rest often reflect substantive theoretical commitments or methodological approaches. Cladistics, for example, prioritizes the descent relationship above all other forms of biological relationships while phenetics tries to remain non-committal. Taxonomies both influence, and are in turn affected by, the broader theoretical commitments of scientists. As with psychiatric conditions, taxonomies can sometimes be affected by social concerns about the consequences of classification. From these examples we see that biologists find order in nature but that the order that they find is often complicated and influenced by their own interests and objectives.

Explaining in terms of mechanisms

One most expects to find order in science in the explanations it offers for the phenomena within its purview. Indeed, the explanatory pursuits of biologists have revealed a degree of order sufficient to ground robust scientific inquiry. But that order is very different than suggested by the popular picture of Newtonian science in which a few simple principles are portrayed as accounting for a wide range of phenomena. Rather, the order is more like that found in the design of human artifacts, where initially simple designs for accomplishing a desired result are modified over time until extremely baroque and not easily understood designs result. A corollary of this process is that initial reductionist aspirations to explain biological phenomena in terms of chemical and physical processes are complemented by much more holistic perspectives that focus on how physical and chemical constituents are integrated into systems as a result of evolutionary modifications.

The term *law* seldom figures in explanations in biology; instead, biologists frequently appeal to mechanisms when offering explanations. The conception of mechanism at work in biology grew out of the ideas of the mechanical philosophy of Descartes and Boyle but quickly expanded beyond the limitation of push-pull interaction of component parts to include processes such as chemical reactivity. Recently a number of philosophers of science whose focus is on biology have tried to articulate the key ingredients in mechanistic explanation in biology. Their accounts of mechanism have paralleled the developments in biology. Scientists begin by proposing what we call *basic mechanistic explanation* which emphasizes the identification of the mechanism responsible for a given phenomenon and decomposition of the mechanism into its component parts and operations. To account for the phenomenon, though, researchers must also recompose the mechanism to show that the parts and operations together can produce it. If the organization is sequential, researchers can mentally rehearse the functioning of the mechanism. However, in the course of further investigation (initiated, for example, to overcome inadequacies detected in the initially proposed mechanism) researchers often find that the mechanism they are investigating employs non-sequential organization. To understand the functioning of such mechanisms researchers must supplement basic mechanistic explanations with computational modelling of their dynamic behaviour, resulting in *dynamic mechanistic explanations*. In either case, the decomposition is followed by an attempt to understand the part in the context of the whole. Surprisingly, biological explanation requires adopting both a reductionist and a holist perspective.

We begin with the reductionistic aspect of mechanistic explanation. Human-made machines require appropriate parts, and a major task in explaining a biological phenomenon is to identify the appropriate parts and the operations they perform. Since biologists are not designing mechanisms but discovering them, they require strategies for decomposing the mechanism they are investigating into their parts and operations. Typically they are not obvious; the scientist must develop ways to intervene on the mechanism so as to reveal the parts and operations. To illustrate this, consider the phenomenon, fundamental to all living organisms, of building their bodies out of nutrients they take in. One important type of part they must construct are the multitude of proteins that facilitate the many chemical reactions that occur in organisms. Uncovering the components of the responsible mechanism was a major accomplishment of the 1950s and 1960s, a project set in motion by the discovery that genes consist of DNA and that particular sequences of DNA specify the individual amino acids that are sequentially linked together to make proteins. The explanatory challenge was to identify the components of the mechanism that mediated between a sequence of DNA and the protein. These turned out to be different forms of RNA each of which was shown to perform a different operation. One can understand the basics of how the mechanism works by following the sequence of operations shown in Figure 4.2. First, as shown in the upper left, an RNA polymerase separates the two strands of DNA and transcribes the sequence of nucleic acids on one strand into a complementary strand of messenger or mRNA. The mRNA is then transported to the ribosomes in the cytoplasm, shown in the lower part of the figure. The ribosomes are composed of another type of RNA, ribosomal or rRNA, that forms a structure on which yet a third type of RNA, transfer or tRNA, can dock to a triplet of nucleic acids on the mRNA. The tRNA consists of a unit that can combine to the three-nucleic acid sequence of mRNA and another that binds the corresponding amino acid. As the tRNA docks on the mRNA, the amino acid it ferries is added to a string of amino acids that

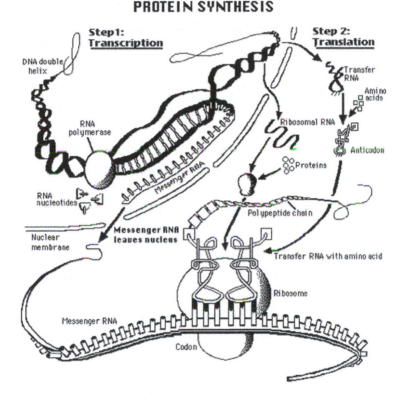

Figure 4.2. Basic steps in protein synthesis. Transcription of a segment of DNA into mRNA occurs in the nucleus. After the mRNA is transported to the ribosome in the cytoplasm, tRNA brings amino acids to dock at successive sites; the amino acids are then strung together into polypeptide chains that constitute the proteins. Figure from Chapter 4: Multiple Orders in Biology, *Genentech's Access Excellence*, Eric Martin and William Bechtel.

constitutes the protein. The mRNA sequence is thus translated into a protein.

This account of the mechanism of protein synthesis is characteristic of mechanistic explanations in that it identifies a number of parts and the operations they perform and specifies how they are organized so that the performance of the different operations generates the phenomenon (the synthesis of the amino acid sequence that constitutes the protein specified in the DNA). Such an explanation is reductionistic in that it explains the phenomenon in terms of component parts and operations of a mechanism. We characterize such an explanation as basic because the organization is limited to the spatial location of the components and the sequence of operations we described. While identifying such mechanisms is an important accomplishment in science, often research on the mechanism reveals that the organization is not sequential and components interact in non-linear fashion. A linear interaction can be characterized using only equations that sum terms for different components and so can be graphed as a line. Any more complex relation, such as multiplication or raising a number to a power, is not linear. Determining the behaviour of non-sequentially organized mechanisms of non-linearly characterized operations typically requires developing an appropriate mathematical representation of the mechanism that can be employed in a computer simulation to determine how the mechanism will behave. The resulting explanations are what we call *dynamic mechanistic explanations*.

Further research on protein synthesis reveals non-sequential features of the way the overall mechanism is organized. For example, whether a given gene is transcribed into mRNA depends on whether the appropriate transcription factor binds to the promoter region of DNA just upstream of the part that is transcribed. Transcription factors are other proteins that are generated from other portions of the DNA so that individual genes modulate the expression of other genes. The interaction of transcription factors often involves complex networks. Even in the bacterium *E. coli*, among the simplest organisms currently living, the network through which transcription factors interact is enormously complex. Figure 4.3 shows about 20 per cent of the interconnections between transcription factors in the bacterium *E. coli* – individual transcription factors are shown as circles and the existence of a line between two transcription factors indicates that one modulates the other.

Figure 4.3 is an exemplar of the representations of complex networks that increasingly are portrayed in biological texts. A cursory examination makes clear that one cannot simply follow a sequence of

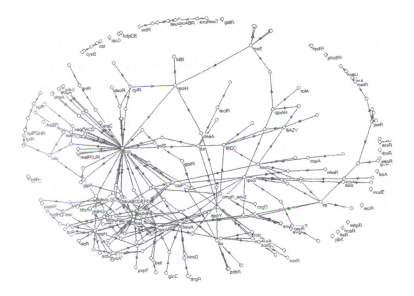

Figure 4.3. Network showing about 20 per cent of the interactions in the transcription network of *E. coli*. Transcription factors are shown as open circles and lines with arrows reflect the action of one transcription factor on another. Figure by Professor Uri Alon, who kindly has provided permission to republish it here. Figure from Chapter 4: Multiple Orders in Biology, *Genentech's Access Excellence*, Eric Martin and William Bechtel.

reactions to understand how the whole mechanism operates. Figure 4.3 and its ilk seem rather disordered. There are, however, patterns of organization or motifs among small numbers of units in Figure 4.3 that can be explained relatively intuitively (although a fuller understanding requires mathematical analysis). Consider the feed-forward loop motif shown in Figure 4.4 in which there is one input from signal S to the mechanism for expressing gene X. There are two pathways from X to Z, one direct and one through Y. Assume the protein produced by X enhances expression of genes Y and Z, and the protein produced by Y also enhances Z. As long as expressing each gene takes approximately the same amount of time, and enhancing expression of Z requires inputs from both X and Y, the result is a network that delays production of Z until S has been present for sufficient time for Y to accumulate. This is useful if, as is common in biological systems, there are random fluctuations in S which would, on their own, randomly generate some unnecessary product. The motif serves as a persistence detector that prevents wasteful synthesis of proteins (an energetically expensive process). Considerable recent research has been directed at identifying such motifs and deter-mining how they behave.

An organizational motif that is common and important in many biological networks is a feedback loop such as that shown in Figure 4.5, in which a gene is transcribed into its mRNA and translated into its protein, which then inhibits the initial transcription process. This arrangement is known as a *transcription–translation feedback loop*. While on first appearances this may appear to be a strange form of organization (one might wonder why a protein should be synthesized just to inhibit its own synthesis), it has the important property of being able to generate oscillations: For a period the concentration of the protein will increase, but once its concentration reaches a sufficient level, it will block its own transcription and its concentra-tions will decrease until enough Z is broken down to allow X to

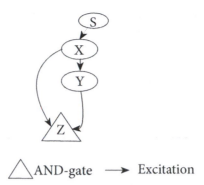

Figure 4.4. A simple organizational motif among transcription factors represented in Figure 4.3 which serves to generate Z only when the input signal S to X is persistent. Figure from Chapter 4: Multiple Orders in Biology, *Genentech's Access Excellence*, Eric Martin and William Bechtel.

begin to transcribe more Z again. Under appropriate conditions, a transcription-translation feedback loop can maintain oscillations indefinitely (provided a sufficient source of energy is available). In biology, oscillatory mechanisms are often employed to segregate in time activities that interfere with each other. One of the best known is the oscillatory mechanism involved in circadian rhythms. Just as we use external clocks to distinguish times for work, meals and pleasure activities, organisms employ clock mechanisms built from feedback loops to temporally segregate the activities they must perform, some of which are incompatible with others. We experience the power of these mechanisms when they are disrupted as in the phenomenon of jet lag – one consequence of jet lag is increased susceptibility to diseases, which results from our immune system providing protection at a time when exposure to pathogens was greatest at our previous location but not in our new time zone.

We have offered two examples, both relatively simple, in which the non-sequential organization of the mechanism results in more

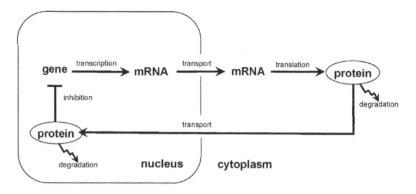

Figure 4.5. Schema of a transcription-translation feedback loop in which a
   protein synthesized from a gene serves to inhibit its own transcription
   until it degrades sufficiently to allow synthesis to begin again. Figure
   from Chapter 4: Multiple Orders in Biology, *Genentech's Access
   Excellence*, Eric Martin and William Bechtel.

complex behaviour than possible with sequential organization. In
both cases, a detailed understanding of how systems organized in
these ways require mathematical analysis (for example, sustained
oscillations in feedback systems require that the component operations be non-linear in appropriate ways) and hence dynamic
mechanistic explanation. We conclude this section with reflections on the implications of the increasing prevalence of dynamic
mechanistic explanations in biology for order in biology. The reductionist perspective of basic mechanistic explanation sought order by
identifying the components of a mechanism and showing how the
phenomenon of interest resulted from the sequential operation of
these components. The recomposition of the decomposed mechanism
required little more than the sequential rehearsal in the scientist's
mind of the component operations. But the discovery that biological
mechanisms often violate sequential order and that the operation of
individual parts is affected by their place in complex networks requires
scientists to focus on how whole mechanisms are organized and how
they are situated in a larger environment that manifests similarly

complex organization. Researchers have often found that networks exhibit *small-world organization* in which, although there are local modules of components, there are also short pathways through which activity in different modules can affect activity within each. In recomposing the results of efforts at decomposing mechanisms, biologists must bring a holistic perspective to complement the reductionistic perspective from which they began. The resulting order provided by biological mechanisms is far more baroque than that to which biologists aspired – the emerging picture is one in which there is not only a different mechanism for each phenomenon, but each mechanism is affected by all of the others in the cell, organism or ecosystem in which it functions. To explain biological phenomena, biologists must be reductionists and holists too.

Descent with modification

The theoretical biologist Theodosious Dobzhansky famously remarked: 'Nothing in biology makes sense except in light of evolution'. One way evolutionary theory helps make sense of biological phenomena is by viewing current life forms as the product of descent with modification. Through much of the nineteenth century, many scientists viewed Darwin's main accomplishment not as establishing natural selection, but as establishing that today's organisms descended from earlier ones. This has provided a crucial ordering principle for many biologists. But as we show, it rests to a large extent on the assumption of vertical inheritance: passing traits from one generation to another. That assumption is now being substantially hedged due to the new-found appreciation of horizontal inheritance, the passing of traits between organisms from different lineages, which could render descent with modification as merely one part of the biological toolkit, rather than a general principle.

Prior to Darwin, comparative anatomists focused on structural similarities between organs in different organisms, dubbed *homologies*. An important clue to whether two traits were homologues was whether they appeared alike in early stages of development. Traits that developed early, scientists reasoned, were more likely to be general characteristics shared amongst organisms of a given kind, with specific differences arising later in development. With the adoption of the evolutionary perspective of descent with modification, traits were construed as homologous when they arose from the same trait in a common ancestor.

The scope of application for homology has since broadened: processes and functions can be homologized, and in the wake of molecular biology, genes and proteins are homologized as well. After genes were identified as sequences of DNA, geneticists and molecular biologists characterized proteins as homologous when the DNA sequences that encoded them were descendent from a common ancestor. Such sequences are also referred to as *conserved*, and many of the basic genes involved in either metabolism or gene expression appear to have been conserved from bacteria and archaea to mammals, including humans.

An example of conservation is found among the peroxiredoxin enzymes that evolved in archaea and bacteria to bind and use oxygen. Following the advent of photosynthesis on Earth about 3.5 billion years ago and the accumulation of atmospheric oxygen it produced, it was highly advantageous to be able to bind this reactive oxygen molecule, essentially a poison because of its ability to disrupt other cellular processes. The amino acid sequence for peroxiredoxins (Fig. 4.6) is virtually the same for many organisms from bacteria to humans, with a few differences in regions beyond the active site. These enzymes have been so strongly conserved presumably due to their usefulness for all organisms in oxygenated environments.

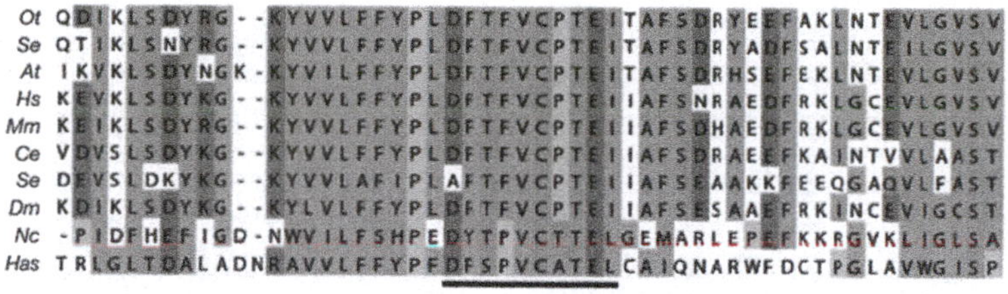

Figure 4.6. Peroxiredoxin amino acid sequence map. The active site is shown by the black bar, on bottom. Sequences are shown for several eukaryotic species (At, *A. thaliana*; Ce, *Caenorhabditis elegans*; Dm, *D. melanogaster*; Hs, *Homo sapiens*; Mm, *M. musculus*; Nc, *N. crassa*; Ot, *O. tauri*; Sc, *S. cerevisiae*) and one species each of bacteria (Se, *S. elongatus* sp. PCC7942) and archaea (Has, *H. salinarum* sp. NRC-1). Reprinted with permission from Macmillan Publishers, Ltd: *Nature* (Edgar, Green, Zhao, van Ooijen, Olmedo, Qin, Xu, Pan, Valekunja, Feeney, Maywood, Hastings, Baliga, Merrow, Millar, Johnson, Kyriacou, O'Neill & Reddy), copyright 2012. Figure from Chapter 2.2: Multiple Orders in Biology, *Genentech's Access Excellence*, Eric Martin and William Bechtel.

One of the major reasons biologists are interested in homologies is that they undergird a powerful research strategy – conducting research on one organism to understand another. For example, much biomedical research ultimately directed at humans is actually done on rodents. The reasons for this are not just ethical – researchers prefer to study a mechanism in the simplest organism in which it appears. But this requires a standard for determining when a studied organism possesses the same mechanism, and homology provides this standard.

Descent thus provides an important ordering principle. This process of descent and diversification was captured in the representations of a tree of life, illustrated in the only figure Darwin included in the *Origin of Species* (Fig. 4.7, lower part): each species is shown as a node on a branch splitting from one root. But the traditional account of vertical descent from parents to offspring, which underlies the tree metaphor and in some cases the tracing of homologous traits back to common ancestors, is being increasingly subjected to challenge.

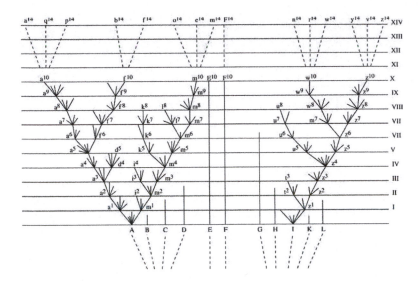

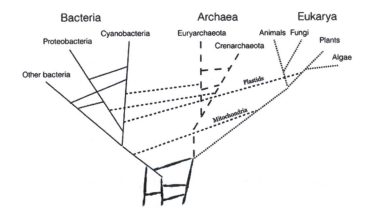

Figure 4.7. Representation of descent with modification in tree diagrams. Darwin's representation of the tree of life (top) shows all descent as vertical – new species split from existing ones. Revised tree (bottom) captures the suggestion of abundant horizontal gene transfer. Figure from Chapter 4: Multiple Orders in Biology, *Genentech's Access Excellence*, Eric Martin and William Bechtel.

On the traditional view, each organism receives DNA just from its parents and passes it on to its offspring; DNA is not shared between organisms. In the mid-twentieth century, though, research on microbial drug resistance showed that bacteria could pass genetic material between themselves and other organisms, not only to offspring, a phenomenon known as horizontal or lateral gene transfer.

Although initially a curiosity, horizontal gene transfer has become recognized as a major factor in the ability of bacteria to resist antibiotics. If one species of bacteria in an environment has a mechanism for combating a drug, it is quickly transferred to those on which the drug was targeted. And this phenomenon is not limited just to drug resistance – the more scientists look, the more horizontal gene transfer they find. In fact, it is not limited even to the world of bacteria. The ability of plants to photosynthesize is thought to be the result of an ancient fusion of photosynthetic bacteria with larger and more complex cells, which together made the first plant cells. Likewise, the ability of your body cells to produce energy, it is now hypothesized, is the result of ancient bacteria adapting to life within larger cells, and over time the bacterial portion developed into what we now know as mitochondria, the crucial 'power plants' within cells. These two cases of *endosymbiosis* are another kind of horizontal transfer, insofar as genetic information was not strictly confined to parent–offspring relations, but actually became part of the larger organism. Horizontal gene transfer is now recognized as a major factor in the evolution of every branch of life. Moreover, the mechanisms that facilitate horizontal gene transfer enable organisms to perform a variety of operations on their own genome, moving genes to locations that suppress or increase their expression and recombining genes in ways that support the creation of novel proteins out of existing components, processes that further challenge the ability to understand conservation as a relatively simple process of descent with modification.

There is some dispute about the frequency of horizontal transfer, and the degree of challenge it presents to the traditional tree of life. Some argue that it is sufficiently rampant that the tree of life metaphor should be rejected; others argue that it adds a few cobwebs to the tree but that the basic tree structure can still be deciphered. We do not take a stand on whether horizontal gene transfer is devastating for reconstructing the tree of life, but rather emphasize that at the very least it complicates the analysis of descent and the attempts to understand the conservation of mechanisms. A mechanism may be conserved either from a common ancestor in a vertical tree or from an unrelated organism in a different phylum.

Conclusion

We have identified three types of order biologists have found in nature, but none of these yields the neat, ordered picture that many have hoped for. In particular, simple laws with wide explanatory power do not seem to be in the offing. While some early evolutionists, including Darwin himself, initially referred to natural selection as a 'law', that label is no longer commonplace. Darwin seems to have conceived natural selection as a universal force acting everywhere at all times. On that view, it would be like some novel Newtonian force, always present in the same way even if its effects varied between individuals. Today natural selection is conceived more as a long-term statistical summary of many disconnected individual facts relating organisms' variation, reproduction and environments. Natural selection is typically referred to as the 'mechanism' of evolution, but not necessarily in the same sense as the mechanistic explanations discussed above.

There is good reason not to expect in biology the sort of universal order that is putatively provided by Newtonian laws. Living organisms have only been identified in a very small portion of the physical world where they appear to have descended through a process of descent with modification in specific environments (even if the process is not totally vertical). Moreover, even if life is found elsewhere in the universe, it is likely, just as life on Earth, to reflect its historical development in a particular location. Whatever biological order is to be found should be expected to be piecemeal and local compared with that provided by any all-encompassing laws. Lacking general laws that might organize the discipline, biology nonetheless employs a number of ordering principles such as the three we have identified. On first appearances, each seems to provide a fairly straightforward mode of order. But in each case we have seen that the order is more complex than first thought.

In retrospect, it often seems easy to identify the oversimplifications of earlier research programmes. Descartes's proposed philosophy of biology appears quite crude to modern biologists. Yet, life sciences evince patterns of inquiry in which initially simpler theories are used to investigate phenomena, and which then give way to increasingly rich details, until the original theory is perhaps no longer even explicitly upheld. For example, early organism taxonomies involved comparing general similarities, which later yielded a multiplicity of dimensions of comparison. Or alternatively, the attempt to decompose biological mechanisms is often attended with dangers, which were sometimes astutely pointed out by nineteenth-century vitalists, but a combination of reductive and holistic scientific strategies eventually supplanted many of the initially simplistic mechanisms proposed. This pattern shows little sign of ending or final culmination: while some physicists search for a 'final theory' which will unify and make sense of the myriad phenomena they study, there is slim hope of such a universal explanation in the life sciences.

References

Alon, U. (2007), *An Introduction to Systems Biology: Design Principles of Biological Circuits* (Boca Raton, FL: Chapman & Hall/CRC).

Dayhoff, M. O., and Eck, R. V. (1965–72), *Atlas of Protein Sequence and Structure* (Silver Spring, MD: National Biomedical Research Foundation).

Edgar, R. S., Green, E. W., Zhao, Y., van Ooijen, G., Olmedo, M., Qin, X., Xu, Y., Pan, M., Valekunja, U. K., Feeney, K. A., Maywood, E. S., Hastings, M. H., Baliga, N. S., Merrow, M., Millar, A. J., Johnson, C. H., Kyriacou, C. P., O'Neill, J. S. and Reddy, A. B. (2012), 'Peroxiredoxins are Conserved Markers of Circadian rhythms', *Nature* 485: 459–64.

Foster, R. G. (2004), *Rhythms of Life: The Biological Clocks that Control the Daily Lives of Every Living Thing* (New Haven: Yale University Press).

Strogatz, S. H. (2003), *SYNC: How Order Emerges from Chaos in the Universe, Nature, and Daily life* (New York: Theia).

5

Realism, Pluralism and Naturalism in Biology[1]

John Dupré

Editorial Link: John Dupré takes this argument from biological practice further by investigating the role that models play in biology. Models are, he points out, both partial and abstractive, and a plurality of models is helpful in understanding biological processes. What is often overlooked in reductive accounts of biological method is the need to take both the wider biological contexts and the developmental nature of biological processes into account. These cannot be captured by a reductive physicalist or 'universal law' type of account, though Dupré insists that this does not licence any sort of 'supernaturalist' entities in biology. KW

Introduction

Questions of realism take on a rather different form in the philosophy of biology from their more traditional context in the philosophy of physics. Problems arise not so much in relation to the existence of the entities that biologists discuss, but in terms of the partial, idealized or abstracted models that they use to describe them. Given this partiality of representation, are we entitled to take such models as true descriptions of the world? There is a burgeoning literature on models in the life sciences, some of which will be assessed in this chapter. I shall argue for a qualified realism about the main classes of such models. However, it is important to recognize that biological models aim at the truth, but not the whole truth. This is why, which is my second main thesis, we should always expect multiple models in biology, providing partial and sometimes complementary perspectives on the reality they aim to illuminate. This pluralistic realism provides a framework for understanding a wide range of issues in the philosophy of biology. Two that I shall briefly discuss are the ineliminable role of context in understanding the nature of a biological entity, and the fundamentally processual nature of living systems. This chapter concludes with a brief remark on the relation of realism to the wider issue of naturalism.

Realism and models

Realism in philosophy of biology poses a quite different question from that traditionally addressed in the context of physics. The question whether belief in quarks, neutrinos and suchlike is warranted by the successful application of theories that employ them remains a contentious one. Bas van Fraassen (1980) is one prominent philosopher of science who has continued to argue that belief in unobservable entities is unwarranted. This does, however, raise a very tricky problem of deciding exactly what constitutes the observation of an entity, and probably more philosophers now would take a demarcation between the unproblematically real and the philosophically debatable, as suggested by Ian Hacking's famous remark, 'If you spray electrons, then they are real' (Hacking 1983: 24). If we can do things with, or to, an entity, then we are entitled to believe it exists.

From this point of view there should be no serious debate about the reality of nucleic acids, amino acids and so on. It is not that they are 'directly' observable: though we can observe these large molecules in one sense through high-powered microscopes, we cannot observe that they are nucleic acids or polypeptides any more than we can 'directly' observe that a track in a cloud chamber is an electron. But we can do lots of things with DNA molecules. Just as one example, transplantation of alien genes into organisms is often ineffective because the host organism preferentially uses different versions of synonymous codons. To address this, scientists can rewrite the transplanted DNA using the preferred codons, and this often enables the transplanted gene to be expressed (Gustaffson et al. 2004). If you can rewrite it, it is real.

Interesting questions of realism arise in biology not about the existence of entities but about the extent to which representations of biological entities, or biological models, correspond to the world. There is a parallel here with a somewhat different traditional question, whether the laws or theories postulated by science are true, a negative answer to which question is often motivated by the so-called pessimistic induction on the history of science, the claim that all past theories have turned out to be false, so ours probably will too. The problem does not arise in quite this form for biology, because most philosophers of biology are sceptical of whether there are any laws or theories in biology, and have described biological ideas in terms of models that are held to represent their target phenomena. Whereas laws have traditionally been taken to be quite literally true (or false), the relation between models and their intended targets is at least more complex.

In what follows I shall refer freely to biological models. However I need to acknowledge that my use of the term is somewhat cavalier in relation to the extensive and often subtle discussion of this topic in recent philosophy. Michael Weisberg (2007) distinguishes scientific modelling from 'abstract direct representation' (ADR). The former involves exploration of a structure that is in some way analogous to the system it is supposed to illuminate, but the elements of which do not claim to represent anything in reality. Many models in evolutionary theory or population ecology are of this kind. Somewhat similarly, Tarja Knuuttila (2005), focusing especially on economic models, holds that scientific models are best seen as cognitive tools, designed to explore a structure taken to be similar in some respects to a target system in the world, but by analogy rather than any direct correspondence. Abstract direct representation, on the other hand, involves representation of specific entities that are supposed to exist, but which are described in ways that abstract from some of their features that are taken to be unimportant given the purposes for which the representations are to be used. In fact, according to Weisberg, 'theorists who practice ADR typically aim to give complete representations' (Weisberg 2007: 229), though they never fully achieve this goal. I am slightly sceptical of this distinction, at least in that I suspect that many cases may be difficult to allocate unequivocally to one category or another. But given that my present topic is realism, my interest is in ADR, or in strategies well to that end of the spectrum, rather than what Weisberg refers to as models *stricto sensu*. As will become clear, I am also sceptical about the goal of complete representation, which may contribute to this partial disagreement with Weisberg.

Biological models in my broad sense are, at any rate, a motley crew. Some biological models are concrete entities, like the highly standardized model organisms that are taken to represent, more or less, much wider classes of organisms (Leonelli and Ankeny 2013), or the sticks and balls arranged to form representations of molecular structures. Others are sets of equations, used to represent phenomena as diverse as the relations of population numbers in ecological systems or the concentrations of chemical species in metabolic processes. Systems biologists try, sometimes, to integrate such mathematical representations into models of much wider biological systems. Structural models may be used to explore, perhaps using computer technologies, interactions between biological molecules. And so on. The models that will be my concern here will be biological models that it is at least strongly tempting to interpret realistically, for example structural models of molecules, metabolic pathways or cells.

What is common to all these kinds of models is that, in contrast to the dominant traditional interpretations of scientific laws, they are not expected or intended to be strictly true. Just as a map will emphasize particular features of the environment and ignore others, on pain of the fate immortalized in Borges's story of the perfect, and thus completely useless, map (Borges 1973), models focus on particular features of the phenomenon they aim to represent, and ignore others. This suggests a simple answer to the question of whether a model is true: if the features it aims to represent correspond exactly with the features in the system or entity represented, then the model should count as true. Unfortunately, models do not typically come with a definitive statement of their intended application, so this criterion is not readily applicable. Indeed, models are often used to explore phenomena by determining empirically the limits of their application, a project that would be rendered unintelligible if the intended scope of application were given in advance.

If science, or at any rate biology, does indeed provide us only with partial and abstracted models of phenomena, are we entitled to claim that they are realistic? As noted above, some models are not intended to be realistic. A paradigm area of science in which this is plausible is neoclassical economics (Knuuttila 2005). But there is an obvious and revealing contrast between economics and models in a field such as molecular biology. No one doubts that there are real economic agents, and no one supposes that there are perfectly rational agents. The object of models involving the latter is not to identify what kinds of entities or processes there are in the world, but to propose a concept in terms of which the aggregation of the behaviours of real agents may be better understood. Whether the model is successful depends on whether typically, or in aggregate, the behaviour of real agents approximates to that of the ideal agents in the model. If it does, then the model may be a useful device for predicting or explaining economic outcomes in the real world. But it does not seem that there is any point at which such a model is intended to correspond to the world.

Compare a model of, say, the citric acid or Krebs cycle, a set of chemical reactions central to metabolism for all aerobic organisms. We do not start here with an uncontested entity that we want to understand better, but rather we postulate a process involving a set of (real) entities whose interactions may help to explain a wide range of empirical phenomena. Either before or after formulating the model we will hope to confirm that these entities do in fact occur in the cells to which the model is intended to apply. We then may confirm, most likely in vitro, that they react together in the required ways. Gradually we gain increasing confidence that we have identified something that really happens in living systems. The purpose of the model is to represent. To believe that it succeeds is to interpret it realistically. It is necessary to bear in mind, however, that a realistic interpretation of the Krebs cycle model doesn't require that something identical to the reactions occurring in a test tube happens in the living cell, nor that the descriptions of entities in the model are complete descriptions of the entities in the world. In vivo, this metabolic process is interconnected with multiple other processes that provide the metabolites and employ the products of these particular reactions. The rate of the reactions isolated in this particular cycle will depend on these contextual processes as will, thereby, its impact on the larger systems within which it is embedded. The model aims to represent a part or an aspect of what happens, but not to correspond exactly with anything real.

Pluralism

This incompleteness of models leads me to my first main thesis: we should welcome and expect a plurality of models of biological phenomena. If we imagine that science gives us the whole truth, we are likely to suppose that the ultimate objective is a single model that represents everything about a phenomenon of interest, as the classical picture described in Chapter 2 suggests. But once we see that models abstract particular features of interest or of relevance to particular questions from the complexity of natural (especially living) systems, this ideal of completeness can be seen to be quite misguided. Different interests will call for different models; and for phenomena of high complexity multiple models will be desirable regardless of any specific questions or interests.

Let me illustrate the pluralism I have in mind with the example of the genome. Nobody now doubts that genomes exist, though at an earlier stage of enquiry successful models of the genome were instrumental in their movement from hypothetical entities to unquestioned existents. Indeed, the possibility of providing multiple consistent though independent models of the genome is now a central part of what puts their existence beyond question (Barwich 2013). Consider, then, some of the models of genomes commonly employed. Most familiar are the sequences of four letters, representing the nucleotides Adenine, Cytosine, Guanine and Thymine. These sequences, we suppose with good empirical reason, often correctly represent the order in which the nucleotides appear in a real molecule. If they do this, then what they say is true. The representation serves to predict or explain, among other things, the subsequent formation of strands of RNA and of polypeptides, with specific sequences of molecular constituents related to that of the DNA.

A quite different representation of the genome presents the general structure of the DNA polymer and its double helical structure, displaying the various chemical bonds that hold the structure together, including for instance the weak bonds between base pairs that are broken when the two strands separate. Such a model, often in the form of a concrete three-dimensional model, is typically used to instruct students about the structure and function of the DNA molecule. It would be possible simultaneously to represent this molecular structure and a specific sequence, though, given the different functions that these models serve, it is hard to imagine any occasion for doing so.

The above three-dimensional structure of the DNA, even if, for whatever strange reason, extended to billions of base pairs' length with a sequence of nucleotides mapping some actual genome (or better now, chromosome, since most genomes contain a number of these subunits), is still far from a complete representation of the genome (or even single chromosome). Actual genomes contain much more than DNA. First we should include the histones, proteins that form the structural core of the chromosome, and around which they are densely spooled, as is required to enable two metres of DNA (in the human case) to fit into a few tens of micrometres of cell diameter. Second, many molecules are, at any time, attached to the genome in ways that crucially determine its behaviour. Various of these bond to particular parts of the genome, inhibiting or enhancing the expression of particular sequences. Of special interest are the small molecules that modify either the nucleotides themselves, or the histone core, in the various processes generally referred to as epigenetics. These modifications change the chemical composition and shape of the chromosome and, by thereby exposing or restricting access of particular sequences to the transcription machinery, help to determine the functioning of the genome.

This last point draws attention to a very fundamental issue: the genome is not a static object, but highly dynamic. Its shape and its behaviour are constantly changing. All of the representations I have just sketched, as indeed any model that represents a structural feature of the genome, are in fact snapshots, frozen in time, of an instant of this genomic process. To imagine a 'complete' model of the genome, it would not be sufficient to somehow integrate all of these various perspectives in three dimensions, but one would need also to extend the representation in time and present the dynamics of the system. Some of these dynamic elements form parts of the typical development of the wider cell or organism; others reflect adaptive responses to contingent features of the wider environment. Both these dynamic aspects are vital parts of the function of the whole. I shall return to this question of living dynamics below.

One reason it is important to be clear on these issues is that it is very common to take a partial representation as the whole truth, or at least the sufficient truth, about some object of enquiry. A striking example of the problems this can cause was the genetic determinism widely propagated in the course of the human genome project. It is possible that some of the notorious statements by prominent scientists, describing the human genome as the book of life or the blueprint for the human, were best understood as propaganda for funding of the project. However, these ideas have certainly disseminated into public understanding so that sequences of As, Cs, Gs and Ts have achieved an almost sacred status, one visible, for instance, in certain arguments to block the development of genetically modified foods. Even if, by 'genome' in these contexts, we meant the full, four-dimensional, concrete entity, including all the aspects discussed above, these statements would be hyperbolic. The genome is a fundamentally important element of all known biological systems, but it is still only one of many without which these systems could not be properly understood. To reduce the organism merely to the sequence of nucleotides in the genome, however, is wholly absurd.

It is, incidentally, easy to see the attraction of the fetishization of genome sequence, as it provides an acceptable way to short circuit the problem of development and the reproduction of biological form. The question of how organisms develop to become as similar as they do to their parents is as fundamental and difficult as any in biology, and the idea that the explanation was there from the start, whether as in ancient visions of an animalcule in the head of a sperm, or as information 'encoded' in the genome, is perennially appealing. This attraction is obvious in Richard Dawkins' (1976) appealing but misguided reduction of evolution to genetics: if the development of the organism could be fully encapsulated in genome sequence, it could be ignored from the perspective of genetic models of evolution. Unfortunately life is not that simple.

The moral of all this for present purposes is the necessity of cautious realism. It is always important to remember that scientific models present only partial truths and abstract from a great deal that is important. But it would be as absurd to deny that genome sequence corresponds to some aspect of reality as it would be to suppose that it provided the whole truth about any real entity. Once this simple point is recognized it becomes almost inevitable that we will want multiple models of a real entity, models that address some of the aspects ignored by a single model. Just the same argument could be directed at the stereochemical and epigenetic models of the genome mentioned above. These models reflect an aspect of reality, but only an aspect. Thus it is not only possible to be both a pluralist and a realist about scientific models, but it is clear that we should be both.

It is worth mentioning in passing that I believe that just the same applies to so-called fundamental physics. Theoretical articulations of reductionism generally start with the premise that physics provides models that apply to absolutely everything, so that ultimately we should be able to understand absolutely everything with reference only to these models (or perhaps rather to some set of laws). Physicalism, the widely held view that there is nothing wholly non-physical, does imply that there are aspects of everything that are physical. Organisms obey the law of gravity, for instance. But that the physiological, ecological, etc. properties of organisms also fall within the purview of physical laws or models seems to me a view that reveals a deep misunderstanding of the nature of scientific representation, and certainly should not be implied by a reasonable physicalism. The extrapolation from models developed to understand the behaviour of the simple microphysical systems constructed in laboratories, or of physical aspects of gross mechanical or chemical systems as found, for example, in astronomy, to everything whatever, is as lacking in justification as the extrapolation I have just considered in genomics. This, unfortunately, is an issue the detailed discussion of which is beyond the scope of our book (but see Dupré 1993; Butterfield 2011).

I have explained why actual models used in molecular biology are partial and abstract from the complexities of the real world. On the other hand such models should presumably be consistent with one another, and if so there seems no reason why we should not aspire to produce some summation of partial models that ultimately acquires the status of a complete model. I do not claim that there is a logical impossibility to some four-dimensional representation of every structural feature of some genome, extended through time to represent the dynamic functionality of the whole system. Such things have sometimes been supposed to exist in the mind of God. But even should there turn out to be such a thing, it is not the goal of science to explore the mind of God. More seriously, even such a four-dimensional model would only represent the history of one particular genome subject to a very likely unique set of causal influences. There is no reason to think it would have any general utility. At any rate, such supposedly complete representations play no part in the actual business of doing real science. Yet this is no reason to deny that actual, partial scientific models may correspond to reality in the specific ways that they are intended to do so.

Context and process

In fact, as just noted, even the complete model imagined above would not provide the kind of universal predictive and explanatory import sometimes imagined by advocates of a complete and final science. I want now to discuss two reasons for this. First is the question of context. Even a complete model of the genome in the sense just discussed would not be a sufficient description for all scientific purposes without some specification of context; context partially determines the properties of a biological entity. Second, and closely connected to the first point, I want to re-emphasize the importance of process. I have already noted that a full description of an entity such as a genome must be four-dimensional, extended through time. But a living process is not a self-contained thing, with its trajectory determined by its internal structure. Its persistence is determined in part by its interaction with its environment. The fact that models of biological structure abstract from the dynamic features of the entity represented reinforces the extent to which the character of the entity is dependent on its wider context.

Even if a model were able to represent everything within its intended domain, it must stop somewhere. So, for example, even a 'complete' model of a cell would not include a full description of the surrounding biological or physical conditions. But a cell, or for that matter an organism, is an open system maintained in thermo-dynamic disequilibrium by its interactions with the environment. So a representation of such an entity is inevitably inadequate to a full understanding of its behaviour. This simple point is already sufficient to show that biological representation is only partial. However, there is a deeper point: the external context in which a biological entity is positioned does not only partially determine its behaviour through interaction between entity and context, it may contribute to deter-mining what the entity is. Or so I argue.

Consider protein-coding genes. One quite legitimate representation of such a gene is in terms of sequence. But genes are also identified in terms of function. In their origins in Mendelian genetics, genes were identified in terms of the difference they made to a phenotype – the genes for eye colour or bristle number, for instance, studied in classic research on Drosophila by Thomas Hunt Morgan and his associates. A particular sequence is capable of determining whether a fly has a particular eye colour, but only given a wide range of background conditions in the cells of the fly. In fact the same sequence could occur in a quite different organism, with quite different effects on the organism phenotype.

Identifying genes as being 'for' phenotypic features is in many ways problematic, as has been extensively pointed out by philosophers of biology (see, e.g. Dupré 2012: 105ff; Griffiths and Stotz 2013). But the same problem arises for more proximate functions. It is natural to think of a 'protein-coding' gene as coding for a particular protein. But it is now well known that a gene may code for many, sometimes even thousands of, different proteins as RNA transcripts are rearranged in the process of splicing, or subject to various kinds of post-transcriptional editing. So again, a description of a gene in terms of the protein it codes for is underdetermined by genetic sequence, and is made true, in part, by further conditions of the cellular environment.

Even the protein does not provide a stopping place for this contextual determination of functional descriptions. It is increasingly clear that proteins can display a wide range of different functions depending on their location within the cellular milieu (Jeffery 1999). It is fascinating that this phenomenon is widely labelled as protein 'moonlighting', suggesting that the protein has a proper function, and is doing something different in its spare time, an implication that surely reflects a residual assumption that a biological entity has at least a primary function determined by its structure. But of course, though one function may be primary in the sense that it occurred earlier in evolutionary history, this has no relevance to a current understanding of the way a system works. So the question whether, for example, some entity is a structural protein or an enzyme will often depend on the broader context in which it is located.

The upshot of these examples is that no spatially restricted description of a biological entity is sufficient for a full understanding of what it is, still less of its behaviour. There are no closed biological systems, and their openness involves ways in which their behaviour is partially determined by their context. This issue has an important bearing on discussions of reductionism. The suggestion that the behaviour of a biological system could be explained or predicted in terms of its molecular parts is clearly vulnerable to the kind of problem just presented. Reductionists are liable to respond that we have just assumed too narrow a scope for the reductive base, and a successful reduction will require inclusion of the relevant aspects of the context. But of course the same problem is sure to arise for the elements of the context that are then included, and there is no reason why this process should terminate. Given the increasing implausibility of practical reduction as the posited reducing base expands in this way, at this point the reductionist will usually retreat to supervenience, the idea that the behaviour of the whole is determined to a sufficient extent by the underlying molecular or microphysical reality. At this point I am inclined to say that, as we move to something approaching a thesis of supervenience on the microphysical state of the entire universe, we move towards a vacuity that verges on the meaningless. Certainly such global supervenience is not an idea with empirical implications. But for present purposes it is sufficient to conclude that actual biological models will always and inescapably be partial, and their applicability will be sensitive to features of their context.

As noted earlier, another perspective on the protean character of biological entities can be gained by suggesting that they are generally better understood as processes than as things. A cell, for example, is a process that originates with the bifurcation of a mother cell, and ends in bifurcation or apoptosis. During the period of its existence it will adopt various structural configurations and contribute to various functions. Its existence is determined not by some property or properties that it exhibits as long as it exists, as we normally suppose for a substantial thing, but rather by a variety of processes that sustain its integrity. In fact, everything in biology is dynamic, and an appropriate metaphysics sees the living world as composed not of things, but of processes (Dupré 2012: Chs 4–5). Rather than see processes as involving things undergoing changes, it is better, at least in biology, to understand things as processes stabilized to some degree over relevant time scales. The 'things' in terms of which we often describe processes are processes stabilized on time scales relevant to the process we are considering (Bapteste and Dupré 2012). Given the current state of physics it is plausible to see the world as composed of processes 'all the way down'; but this is an issue that can be left moot for present purposes.

To different degrees, all models abstract from the four-dimensional character of biological processes. Many or most biological models are of course dynamic to the extent that they represent the way a system develops over time, perhaps depending on some external variables. But this is already too linear an approach to capture the full dynamic complexity of the system. As J. S. Haldane (1931: 22, cited in Nicholson and Gawne 2013) put the point: 'Structure and functional relation to environment cannot be separated in the serious scientific study of life, since structure expresses the maintenance of function, and function expresses the maintenance of structure.' I do not want to argue that it is impossible to provide a representation of all this dynamic complexity, but only to insist that there is very likely no reason to do so. Once again, we need only recall the specific purposes for which we represent parts or features of living systems, and the uselessness of the 'perfect' Borgesian map.

Realism and truth

My general point so far has been that biological models are partial, but nonetheless the entities that they refer to are, often enough, real. This does leave an awkward question, however, whether the things they assert are true. If I present a model involving perfectly spherical cells, it may be acceptable to say that I am referring to the actual, not quite spherical, entities in the world. But surely if I say they are spherical, I say something false. Generally, it is impossible to avoid the conclusion that if a model is intended to assert the existence of the states of affairs it portrays, then it will almost fail in this intention.

Considerations of this sort have led many philosophers to deny that statements based on scientific models are generally true. Several responses to this problem are possible. One would be to collapse Weisberg's (2007) distinction between models and abstract direct representation towards the former, and claim that in both cases we are dealing with some kind of cognitive instrument that illuminates the world by analogy rather than by direct description (Knuuttila 2005). Some go further and argue that scientific models are best seen as a kind of fiction and can be illuminated by exploiting ideas about literary fiction (e.g. Frigg 2010; for a general discussion of this strategy see Godfrey-Smith 2009); but this strategy, while it may be defensible for certain classes of models, is surely inappropriate for the kinds of partial representation I have been considering. My own preference is for a pragmatic approach that takes scientific statements as true, or perhaps just correctly assertible, if they correspond with the world in the intended respects. To illustrate with a very simple example, it is true, in appropriate contexts, that glucose is $C_6H_{12}O_6$, and also that fructose is $C_6H_{12}O_6$. No chemist would be tempted to conclude that glucose was identical to fructose, though there might be contexts in which the differences between these sugars were irrelevant. Partial representations licence assertions, but only in a suitably limited class of contexts. The extent to which scientific descriptions are adequate to their intended uses is always a matter open to further investigation. A telling and tragic example is provided by the chemical thalidomide, which normally exists as an equal mixture of two optical isomers, mirror image structures, an apparently very subtle difference. One of these isomers is an effective treatment for morning sickness; the other causes drastic developmental abnormalities. For most purposes the three-dimensional structure, not specifying either of the optical isomers, would be a fully adequate chemical description; for pharmacological purposes it was clearly catastrophically insufficient. Whatever view is taken on truth, the crucial point is that the limitations, known or unknown, of particular scientific perspectives must always be borne in mind.

Naturalism

A wider context in which discussions of realism figure significantly is the general philosophical thesis of naturalism. Unfortunately naturalism is a highly contested concept. My own view is that a more useful thesis is anti-supernaturalism (Dupré 2012: Ch. 1). That is to say, naturalism is the view that a proper ontology should be restricted to entities that have some empirically accessible connection to the natural order; it is a negative thesis rather than a positive one. The objective should be to leave the characterization of the natural order as vague as possible, while allowing enough substance to exclude such entities as deities or immaterial souls. Of course it is not an *a priori* truth that there are no deities or souls. If Jupiter were to appear in public view and begin casting thunderbolts and mating with swans, our evaluation might rapidly change. Theology might become a respectable branch of empirical science. But while gods continue to play no detectable role in public life, a naturalist will exclude them from his or her ontology. The point of the vagueness is to prevent substantive positive theses about the world being smuggled in under cover of this purely negative thesis. Materialism, for example, is sometimes taken to imply no more than the exclusion of the immaterial, where this has very much the same import as the supernatural. But there is also a tendency to equate naturalism with materialism and thence often to reductive physicalism. Concepts such as emergence or so-called downward causation, the determination of the behaviour of parts by wholes of which they are parts, are said to be 'spooky', indicative of the supernatural. On the contrary, these seem to me clearly contingent and empirical hypotheses about how the natural world works.

The motivation for naturalism is often, and appropriately, the view that the broadly empirical methods we think of as scientific have provided an epistemology for exploring the world vastly more successful than the appeal to supernatural agents lying outside the natural order. Our ontology, it is then argued, should be based on successful epistemology. This motivation makes clear why some kind of realism is necessary for such a motivation of naturalism. Without a realistic interpretation of science no link can be made from successful science-based epistemology to ontology. As I hope to have shown, while it is not trivial to provide a realistic account of biological methodology, there are no insuperable obstacles to doing so, and indeed the successes of modern biology require some kind of realistic interpretation.

Note

1   An earlier version of this paper appeared as: 'Réalisme, Pluralisme et Naturalisme en Biologie', T. Hoquet and F. Merlin (eds), *Précis de*

Philosophie de la biologie (trans. T. Hoquet), Paris: Vuibert, 2014, 169–82. The research leading to this paper has received funding from the European Research Council under the European Union's Seventh Framework Programme (FP7/2007–2013) / ERC grant agreement no. 324186.

References

Bapteste, E., and Dupré, J. (2013), 'Towards a Processual Microbial Ontology', *Biology and Philosophy* 28: 379–404.

Barnes, B., and Dupré, J. (2008), *Genomes and What to Make of Them* (Chicago: University of Chicago Press).

Barwich, A. (2013), 'Fiction in Science? Exploring the Reality of Theoretical Entities', in G. Jesson, G. Bonino and J. Cumpa (eds), *Defending Realism: Ontological and Epistemological Investigations* (Berlin: De Gruyter Publishers), 291–309.

Beatty, J. (1995), 'The Evolutionary Contingency Thesis', in G. Wolters and J. G. Lennox (eds), *Concepts, Theories, and Rationality in the Biological Sciences: The Second Pittsburgh-Konstanz Colloquium in the Philosophy of Science* (Pittsburgh: University of Pittsburgh Press), 45–81.

Borges, J. (1973), *A Universal History of Infamy*, trans. N. T. di Giovanni (London: Allen Lane).

Butterfield, J. (2011), 'Laws, Causation and Dynamics at different Levels', *Interface Focus* (Royal Society London) 1: 1–14.

Dawkins, R. (1976), *The Selfish Gene* (Oxford: Oxford University Press).

Dupré, J. (1993), *The Disorder of Things: Metaphysical Foundations of the Disunity of Science* (Cambridge, MA: Harvard University Press).

Dupré, J. (2012). *Processes of Life: Essays in the Philosophy of Biology* (Oxford: Oxford University Press).

Fraassen, B. van (1980), *The Scientific Image* (Oxford: Oxford University Press).

Frigg, R. (2010), 'Models and Fiction', *Synthese* 172: 251–68.

Godfrey-Smith, P. (2009), 'Models and Fictions in Science', *Philosophical Studies* 143: 101–16.

Griffiths, P., and Stotz, K. (2013), *Genetics and Philosophy: An Introduction* (Cambridge: Cambridge University Press).

Gustaffson, C., Govindarajan, S., and Minshull, J. (2004), 'Codon Bias and Heterologous Protein Expression', *Trends in Biotechnology* 22: 346–53.

Hacking, I. (1983), *Representing and Intervening* (Cambridge: Cambridge University Press).

Haldane, J. (1931), *The Philosophical Basis of Biology* (London: Hodder and Stoughton).

Jefferey, C. (1999), 'Moonlighting Proteins', *Trends in Biochemical Sciences* 24: 8–11.

Knuuttila, T. (2005), 'Models, Representation, and Mediation', *Philosophy of Science* 72: 1260–71.

Leonelli, S., and Ankeny, R. (2013), 'What Makes a Model Organism?', *Endeavour* 37: 209–12.

Mitchell, S. (2003), *Biological Complexity and Integrative Pluralism* (Cambridge: Cambridge University Press).

Nicholson, D., and Gawne, R. (2013), 'Rethinking Woodger's Legacy in the Philosophy of Biology', *Journal of the History of Biology* 47: 243–92.

Sober, E. (1997), 'Two Outbreaks of Lawlessness in Recent Philosophy of Biology', *Philosophy of Science* 64 (Supplement): S458–67

Weisberg, M. (2007), 'Who is a Modeler?', *British Journal for the Philosophy of Science* 58: 207–33.

6

The Making and Maintenance of Social Order

Eleonora Montuschi and Rom Harré

Editorial Link: The contribution by Rom Harre and Eleanora Montuschi takes us further up the 'hierarchy' of sciences to scientific approaches to the existence of persons in society. Here the emphasis is on how different scientific explanation is when applied to societies from explanation in the purely physical sciences. The emergence of persons adds new dimensions of symbol-using relationships, to understand which we must have a grasp of semantic information, which does not even occur in physics or biology. We must have a grasp of moral evaluations, of the informal systems of rights and duties with which 'positioning theory' in the social sciences is concerned. And we must see how persons are embodied conscious beings (moral entities) embedded in local cultures. There are 'no universal laws of social order'. Thus the existence of conscious beings in evaluative and symbol-using societies gives rise to new forms of order which cannot be reduced to or predicted from physics, chemistry or biology alone. These are emergent, local, historically developing and holistic forms of order which require their own distinctive forms of explanation which are not entirely naturalistic, even less so mechanistic, and thus throw doubt on attempts to make physics the all-explanatory super-science. KW

Social order: The very idea

What is social order? And where do the ideas of order we use to describe social order come from?

One of the oldest suggestions invites us to look at 'the orderliness of nature' as a possible source. The succession of the seasons, the repetition of the forms of animals and plants over the generations, is striking. However, the earliest attempts at an explanation of natural order were driven by the threat of the eruption of disorder – as if, were nature to be left to its own devices, it would disintegrate into unmanageable chaos. This could be prevented only by divine decree, or by human management such as gardening and animal husbandry. When nature is subdued to a rule, for example by divine decrees, it becomes an orderly system. When animals and plants are managed by a farmer it becomes a controlled environment. The Garden of Eden was an orderly place under divine management until Eve and the serpent messed it up and thereafter people had to create order for themselves. Fortunately, Adam had already named all the plants and animals in the Garden!

Is something similarly true of society? If human life is bereft of those constraints that ensure orderly person interactions, what prevails is a 'state of nature'. It was under these conditions that Hobbes declared 'the life of man: nasty, poor, brutish and short' (Hobbes 1651). There would be no orderly polity without an imposition of the power of a sovereign to maintain order. By accepting norms, human life becomes instead regulated by some form or another of a social contract – as in a democracy – or by contrast falling under a dirigiste government such as a theocracy. Whichever way it is created and maintained, social order depends on people following rules and conventions, though these important guiding devices may become visible only when infractions occur, and rules and conventions are needed to display some actions as infractions.

The natural world displays order under certain conditions, so – the oldest suggestions go – society should imitate the orderliness that is displayed in nature.

What does any order worthy of the name, be it natural or social, demand then?

One feature is predictability. Were we not able to project past and present over the future, we would have no guide for action, nor would we know how to adjust means to ends or situate ourselves in the course of events. We would not be able to describe what happens either in nature or in social relations because description requires the use of common, general terms, and these must have more than a one-off application. Our descriptions of natural events and our dealings with each other require some measure of stability of expected responses (or of effects).

A second element is 'organic integration', based on the belief that each part has a functional role in a whole. In order to be fit for its task, each part needs to respect a hierarchy of functions – be it the structure of a beehive, the functioning of a clock or the tax collection system.

A third feature, and one that perhaps best defines the essence of order, is regularity. The idea of universal natural laws, as developed by late-seventeenth-century natural philosophers, best represents this essential aspect of order as regular (i.e. constant, or at least significantly frequent) co-occurrences of similar sets of events.

Newtonian physics and evolutionary biology are two well-known descriptions of the orderliness of nature that were used to view social order. They indeed provided political, religious and economic sources for the creation or confirmation of particular forms of social order. Let's briefly recall how.

Natural/social: The roots of comparison

Modern physics assimilated the universe to a gigantic machine – a cosmic clock. That meant a number of things. First, once set in motion a clock will behave always in the same way (and so will its constitutive parts). Second, to understand the behaviour of a clock we need to understand how the interactions of its parts will give rise to the behaviour of the machine as a whole. Each part is related to the others according to a fixed hierarchy – from the simple to the complex. Third, the fact that something is, say, a clock depends on what its contituent parts are and to what kind they belong. The same goes for the world, if it is thought of as a machine.

The powerful simplicity of Newton's three mechanical laws of motion fired the imagination of several social, political and economics thinkers between the sixteenth and nineteenth centuries to try and discover the basic principles of social life. From Hobbes to the empiricist philosophers John Locke (1691) and John Stuart Mill (1859), the economist Adam Smith (1776) and the psychiatrist Sigmund Freud (1929), in describing the state, society, the market, human history or the presumed mysteries of the unconscious, they all drew from the same source, physics (and chemistry). The father of sociology, Auguste Comte (1861), called his newly born science 'social physics', and the model of physics he had in mind was indeed Newtonian.

By the end of the nineteenth century, Darwinian science was also seen as offering substantial clues to understanding society. Herbert Spencer (1862) conceptualized society as a 'social organism', developing according to the universal laws of evolution and aiming at the 'survival of the fittest' (an expression he coined to describe the leading principle of human relations in social interactions). Social Darwinism, as it came to be associated with the name of Spencer, was both a theory of society and a moral perspective concerning how a good society should function. It had a deep effect on both the ways social institutions were conceived, and the ways public policies were formulated.

Systems of government were conceived as centralized, hierarchical and controlling from the top down (as for example in the so-called British Westminster political model) and policies were conceived in such a way that they could rely on 'rational' procedures in the processes of decision-making. This entailed a series of assumptions: that they addressed stable systems and well-defined problems, that they targeted unitary actors (e.g. the government) and that these actors make (or tend to make) rational choices. Besides, it is assumed that all institutions, all economic interactions, all processes and textures of society, have an end state or final equilibrium towards which each of them converges while the totality of social moieties develops towards an end state. The nineteenth-century German philosopher G. W. F. Hegel (1821) and the contemporary American political scientist Frances Fukuyama represent two versions of this conception. Policies should be conceived with a view to reaching that final stage.

These images, however, also proved their limits when confronted by the large and complex variety of aspects that a description of order in the social realm seems to entail. For example, there is little in systems of communication among animals and plants that could serve as models to inform the very complex and sophisticated systems in use among human beings. Rather the modelling goes the other way – what we have been able to discover about animal and plant communication depends on deploying models derived from human systems. Also, as we will show later in detail, the use of symbols to manage our interpersonal relations and the main means for our individual cognitive practices does not seem to be adequately captured by either a mechanical or gene-led description.

When saying that humans and animals are close, how close is close? For example, to the questions: 'why should people help one another?' and 'who should we help most of all?', should we be happy with the answer that altruistic behaviour is a function of kin selection? That is, we tend, like any other animal, to favour kin over non-kin, and to favour close kin over distant kin (in ways that can be calculated in terms of percentages – 50 per cent interest in its parents,

offspring and full siblings; 25 per cent interest in half siblings, grand-
parents, grandchildren, uncles, aunts, nephews and nieces; 12.5 per
cent interest in first cousins, half nephews, great-grandchildren, etc.).

If this were to be the case, we might still wonder, how can we
explain social altruism, for example altruism among strangers, or
altruistic acts towards people we do not know and will never meet, the
recipients of gift aid and other charities? Why should we help the aged
(or now infertile), if the aim is genetically successful reproduction?
Are kin individuals always 'deserving' individuals (is a great-grand-
child less 'deserving' than a brother simply because the genetic interest
of the giver in the former is half if compared with the latter), etc.?

Even the old school of sociobiology (the early supporters of genetic
altruism) would admit that cultural evolution has produced a wide
variety of modifications on the basic biological model. Though the
sociobiologists continued to argue that the *fundamental* explanation
of social behaviour rests on genes, this still points to the fact that a
genetic-type explanation does not, after all, have the answers needed
in particular circumstances such as the effect of scientific discoveries, linguistic changes, cultural innovations, volcanic eruptions and
climate change. Societies make their own adaptations to these irrup-
tions into a standing social order.

Evolutionary psychology, by shifting attention from behaviour
to the inner constitution and functioning of the human brain, has
been aimed at offering a more satisfactory, causal explanation of why
individuals act in certain ways within certain environmental circumstances. At best it can explain how some people are disposed to act
in certain ways – even when the environment is not conducive of
those features that activate their dispositions. There may be an innate
propensity towards religious belief but it is realized in many sects and
versions, and these may be adhered to when the social circumstances
for their existence have become hostile. How can there be a Coptic
Church in contemporary Muslim Egypt?

However, this recent perspective fails to catch the detail of social order, its creation and maintenance. For example, we might admit, though the claim is disputable, that some individuals have a capacity for rape, a brain 'module' that drives violent acts against women, but this does not explain how such attacks occur in certain specific social contexts, how society reacts to complaints, and so on. In other words, there might be a natural reason that predisposes individuals to act in certain ways, do certain things, but that very reason might be neither sufficient nor relevant to account for what we want to understand: actual people doing specific things in quite definite circumstances. Social behaviour comes in a great variety of forms of social acts, rarely explicable in terms of their biological roots alone.

In this chapter we do not intend to focus on well-rehearsed and widely discussed arguments on how to compare or combine natural science with social science, nor on whether it makes any sense to try to reduce the concepts and methods employed in the latter to the former. Instead we will show how nowadays there are informative and exciting ways to describe and explain order in social domains that do not require borrowing from the concepts, theories and procedures of natural science. One such perspective is that which goes under the label 'evolutionary game theory'. It consists in the application of the theoretical apparatus of game theory to biological contexts, and it has become particularly interesting for sociologists, economists and anthropologists as a way to explain a number of aspects of human behaviour, such as altruism, morality, empathy and social norms. As applications and discussions of this perspective can be easily found in the literature (EGT has indeed become a most fashionable and popular theory of social order), in this chapter we chose to focus on a different, perhaps lesser known, perspective. Coming from social psychology, it has developed a set of tools that do not require arguing analogically from physics to social science, nor reducing social science to genetics. It rather effectively addresses core aspects of social order by means of its own, well-equipped (literal) language of description.

Social order: The study of 'positions'

However influential our genetic make-up may be in the generic forms of human social order, such as the nuclear family, the dominating feature of human life is the use of symbols to manage our interpersonal relations. There are flags, road signs and dress codes, and there are the innumerable ways we talk to each other apropos of how our relations should be. The expression of a local social order takes place largely through the meanings that are understood or assigned to the social icons that are scattered around the environment. Think of the vast statue of Abraham Lincoln that presides over the Washington Mall, and think also of the fate of the statues of Saddam Hussein in the expression of the fracture in social order that occurred during the second Iraq War.

Recent work in social psychology has turned to how beliefs about rights and duties are distributed among a group of people and serve as a key feature of the evolution of patterns of social life, from the intimate to the international (Harré and van Langenhove 1999). The cluster of ways a person, group or nation takes its rights and duties to act in certain ways has been called a 'position'. The study of positions, and how they are established, challenged and implemented, is 'positioning theory'. Positions are not the same as 'roles'. 'Role' is an upgraded commonsense notion – what you need to do to be taken to be a person of a certain sort. What sort? Well, there are roles like 'mother', 'son' and so on which can be qualified as 'good', 'bad' or 'indifferent' depending on local and historically particular conventions. Then there are roles like 'judge', 'doctor', 'priest', 'executioner', 'surveyor' and so on that are the active aspect of recognized professions. Each role includes a pattern of taken-for-granted, although sometimes explicitly formulated, rights and duties with respect to the performance of repertoires of social acts. The killing of someone by an executioner is the consequence of the fulfilment of a duty and the exercise of a right by the institution that has decreed it. Killing someone in a fit of rage, to get the treasure map and so on, is not embedded into a long-standing normative pattern, though at that moment the assassin may believe that he or she has a right to the treasure. The boundary between rightful and dutiful killing is culturally contingent: witness the way that Islamic honour killings are treated as murder in the UK.

A social psychologist might find it useful to study the distribution of rights and duties in relation to the roles that are recognized in a certain culture, eschewing any presumptions about the spillover of such patterns elsewhere and at other times. However, more significant for the understanding of the unfolding of the episodes of everyday life is the informal, often ephemeral distribution of rights and duties among a small group of people in close and even intimate contact. This is indeed the focus of the recent developments of positioning theory in a wide variety and scale of research projects. Who has the right to speak, when and to whom in the course of a conversation? Who has the duty to put out the refuse bins for collection?

There are three relevant background conditions for understanding those episodes as instances of social order. The first background condition is the local repertoire of *admissible social acts,* which are the social meanings given by the local people to what is said and done in any particular episode. The same verbal formula, gesture, flag or whatever may have a variety of meanings depending on who is using it, where and for what. Saying 'I'm sorry' may, in certain circumstances, be an apology. It may also, in the UK, be a way of asking someone to repeat what has just been said. It may be a way of expressing incredulity or even a way to reprimand someone.

The second background condition is the implicit pattern of the *distribution of rights and duties* to make use of items from the local repertoire of socially meaningful acts and utterances. Each assignment of positions creates a distribution of rights and duties. A mother has the right to discipline her child in whatever way laws and customs allow, but a visiting neighbour does not. Catholics have a duty to confess their sins individually, while Protestants do not. Human beings in general are not required to make confessions, only Christians.

The third background feature to social order is the repertoire of *storylines* which are usually taken for granted by local social actors. Autobiographical psychology, the study of how, why and when people 'tell their lives' and to whom, also draws on repertoires of storylines. A train journey may be told as a 'heroic quest', and what would have been complaints about lateness according to one storyline become obstacles to be bravely overcome in another. A solicitous remark can be construed as caring according to one storyline, but as an act of conde-scension according to another. Besides, the orderliness of sequences of events in social episodes is seen in the way stories repeat themselves endlessly. However, there are other ways that order is predetermined. There are customary ways of carrying out everyday tasks that shape and maintain social order but they are too variable and contingent to be assimilated to the psychology of roles. How we eat our food is registered in etiquette books, but mostly 'diner' is not a role.

The three background conditions that are constitutive of a local social order mutually determine one another. Presumptions about rights and duties are involved in fixing the moment-by-moment meanings of speaking and acting, while both influence and are influ-enced by the taken-for-granted storylines realized in an evolving episode. An example will illustrate the value of using positioning theory to analyse the underlying structure of presuppositions that affect the unfolding of an episode.

An example of positioning analysis

A central feature of the way that liberal democracies create and maintain social order is the devices by which an orderly transition from one ruler to another is accomplished at regular intervals without recourse to bloodshed. The pattern of rights and duties that sustain our most valuable form of social order involves a sequence of positioning moves in which rights and duties are contested in debates and elections.

The systems that have evolved in 'Western democracies' involve a sequence of contests by means of which, step by step, candidates acquire the right to govern. However, the process is complex. To begin with, a candidate needs to establish the 'right to stand in the election'. The discourses offered by those who are standing for public office in countries with systems that pit two or more parties against each other involve two related main threads in their presentations. The candidate must make clear that he or she has the right to enter into a context, be it democratic or for decision in some other way, say by a selection committee or board. This is not a matter of positioning, but of conformity to laws enacted by the nation where the contest is to take place. The candidate must also display for those who will choose among the contestants whatever skills, knowledge and 'characterological' assets would make it overwhelmingly wise to select just this person for the post in question, even if it is the right to stand for another post. It also helps that a candidate should at least pretend that he or she feels a duty to stand. This candidature is no whim, nor is he or she who claims it being pushed or forced into standing. Now we enter the realm of positioning theory.

Recent developments in positioning theory have brought to the fore the importance of the pre-positioning processes by which people justify or undermine the rights of themselves or others to take up a post, begin a course of action, claim the possession of something and so on. A formal version of pre-positioning goes on in courts and tribunals concerned with property ownership, for example by inheritance. Who has the right to the bulk of the old man's fortune? If the matter has not been settled by the authority of the will, then positioning disputes will be likely to occur, in the here-and-now establishment of a right. The tycoon's trophy-wife will pre-position herself by asserting that she made his declining years a joy, while his estranged daughter, upset by the advent of the trophy-wife, might insist on ties of blood.

These pre-positionings bear more or less directly on the distri-bution of rights, but it is not for the court to decide the distribution of rights among the relatives. Pre-positionings are notoriously contestable – the daughter calling on the housekeeper to testify to the shrewish tongue of the wife, and the wife calling on the same person to testify to the daughter's neglect of her poor old father. Rights to matters that lie outside the jurisdiction of the courts are contested via the contesting of the pre-positionings just as they are in the formal proceedings of the probate hearings.

The system of primary and general elections in the United States is structurally similar to the format of this kind of dispute, in which conversational exchanges may finally lead to an agreement. We can see this in the primary campaigns of the Democratic party's would-be candidates for the presidency, two aspirants who are now President Obama and former Secretary of State Clinton.

At the beginning of the long process of choosing a party candidate it seemed that Hillary Clinton was, as they say, 'a shoo in'. However, a challenge quickly emerged from a dynamic and charismatic senator, Barack Obama, not then known to the American public. By early 2008 his challenge to Hillary Clinton for the right to be the candidate of the Democratic Party looked increasingly powerful. This led to the proposal of a series of debates in which the leading candidates,

Senators Clinton and Obama, would meet face-to-face to debate the issues of the day. The point of these events was to establish a prior right – the right to be the Democratic candidate. Achieving the right to rule was not a process to be analysed by the use of positioning theory – that right was determined by a formal procedure, the general election.

These debates were ideal exemplars of the use of positioning theory in social psychology. The debates involved the three components of any positioning analysis –the admissible repertoire of social acts, the establishment and distribution of rights and duties, and storylines (Harré and Rossetti 2011).

Both Obama and Clinton drew on autobiographical material as pre-positioning moves. In response to the charge that he criticizes people for 'clinging to their religion', Obama responds with the following: 'I am a devout Christian … I started my work working with churches in the shadow of steel plants that had closed on the south side of Chicago, [I claim] that nobody in a presidential campaign on the Democratic side in recent memory has done more to reach out to the church and talk about what are our obligations religiously …' (CNN Democratic Candidates Compassion Forum, 13 April 2008). Here Obama is making a reactive pre-positioning in response to an implicit positioning move that would undermine his right to be a candidate for the formal right to rule in a Christian country.

Obama laid claim to a special attribute, the ability to bring people together: 'what was most important in my life was learning to take responsibility not only for my own actions, but how I can bring people together to actually have an impact on the world' (ABC News, 26 February 2009).

Responding to a neutral query, Clinton makes a pre-emptive pre-positioning move by introducing an autobiographical snippet: 'You know, I have, ever since I was a little girl, felt the presence of God in my life. And it has been a gift of grace that has for me been

incredibly sustaining. But, really, ever since I was a child, I have felt the
enveloping support and love of God and I have had the experience on
many, many occasions where I felt like the holy spirit was there with
me as I made a journey' (CNN Democratic Candidates Compassion
Forum, 13 April 2008).

There is a striking contrast in how Obama and Clinton discuss
their experiences with religion – Obama addressing churches as
points of community connection and religion as a duty; Clinton
approaching her faith as a matter of personal experience and feelings.
This contrast plays into significant narratives for each candidate,
supporting Obama's self-proclaimed role as a paternal community
organizer and reinforcing Clinton's image of an emotional and strong
maternal character.

In the context of a back-and-forth series of accusations and
responses over the details of their different proposals for a universal
healthcare plan, straightforward accusations of character effects
appear as Clinton and Obama move from a discussion of the merits
and demerits of their schemes to a discussion of the merits and
demerits of their characters.

Clinton needed to disperse the impression that she was a different
kind of person, indeed a superior kind of person, from the bulk of
Democratic voters whose support she was canvassing. She had to
produce a story of her life that showed her in that light. We could call
it the 'Ich bin eine Berliner' storyline that Jack Kennedy deployed
with good effect: I am not some person from a distant place whose
interest in your welfare is merely theoretical – I am one of you and
your interests are my interests.

Obama then uses the autobiographical story quoted above as a
storyline to show that he too is 'just one of you. ... working as a civil
rights attorney and rejecting jobs on Wall Street to fight for those
who are being discriminated against on the job – that cumulative
experience is the judgment I bring' (ABC News, 26 February 2008).

Obama does not have to apologize or explain why he is on the side of the underdog – he was an underdog! At the same time he must also claim that he is not really an underdog – only as a Harvard law graduate does he have the resources to serve the people whom he is addressing. Clinton has to account for her positioning of herself as having a duty to serve the poor and underprivileged – but it is as Lady Bountiful that she is making her claim. She must demonstrate that she is not less sincere than Obama in positioning herself thus – but she must do some pre-positioning digging into her religious history to account for her benevolence. However, in claiming that she has a right and duty to serve the poor, Clinton runs into a narrative tension between privilege and struggle. In the Texas debate, she remarks that 'with all of the challenges that I've had, they are nothing compared to what I see happening in the lives of Americans every single day' (ABC News, 26 February 2008). In one breath, she has tried to assert that her experiences with struggle make here a viable candidate to understand and represent the poor even while recognizing a distance between her silver spoon upbringing and the everyday American. In trying to run both storylines, Clinton jeopardizes the integrity of both.

What is most interesting about positioning theory is that it examines how people, as a matter of fact, distribute rights and duties among the members of a group in all those everyday contexts that lie outside the formal distributions of rights and duties normally enacted by socially acknowledged roles, or by court judgements.

Closer to the interest of this book, the relevance of positions to the establishment and maintenance of social order is very direct. An interlocking network of beliefs about people's rights and duties, with respect to each other and to the public good, provides the psychological foundation for people to live in relative harmony. The psychology in question is moral rather than scientific: it requires the willingness to debate the assignment of rights and duties and to accept that assignment, at least until another occasion for contesting should arise.

Persons as the root concept of any social order

If we are to maintain the ineliminability of moral orders from our understanding of the core conditions for there to be a human society, we must protect the core concept of 'person' from being downgraded into a mere organism the behaviour of which can be explained without remainder by the use of genetics and neuroscience. Human beings *qua* higher animals are 'minded' parts of a natural order. However, to impose the metaphysics of a generic natural order on the cultural practices of human beings is not just a mistake, but a fallacy: that is, a conceptual error. In some cases, especially visible in contemporary 'neuropsychology', the error is a mereological fallacy, that is a case of the fallacy of ascribing, to a part of a person as embodied conscious being, a predicate the meaning of which is determined by its use for the whole person (Bennett and Hacker 2003): for instance, declaring that the hippocampus or the entorhinal cortex 'remembers' or the auditory cortex 'hears'. Remembering is something the whole person does. It is not just to contemplate a recollection of the past, or offer a statement in the past tense as a description of something that once happened. To be an act of remembering, that description must be correct, ceteris paribus, and something to which the speaker has an intimate connection, such as 'being there'. To say 'I remember ...' is a social act and subject to moral as well as empirical judgement. Reporting a recollection in this form entitles another person to repeat the claim with conviction. We are blamed for hasty claims to remember.

How should we qualify 'human being' for the purposes of defining the field of such a suitable psychology? What sort of 'order' is it that encompasses human life as it is lived? Cultural-discursive psychology, exemplified in the writings of Lev Vygotsky (1975), Jerome Bruner (1986) and others is concerned with the attributes of human beings, but not *qua* higher animal, the whole living animal, rather to the human being *qua* person, such as 'reminisces with old friends', 'chooses from the menu', 'hopes for a fine afternoon', 'finds the prisoner "guilty"' and so on. What makes these phrases special is that their uses are embedded in local culture, and what they mean case by case depends in part on the details of that culture. To develop our understanding of the kind of order that is exemplified in the psychological and social activities of human beings, we must elaborate the concept of 'person' in more detail as it appears in the foundations of these studies. We can attempt only a brief sketch of what 'person' means in cultural-discursive psychology and as the core of conceptions of social order. As the concept is used in hybrid psychology, a person is a singularity having a unique trajectory in space-time, embedded in a web of material relations to other persons as embodied beings. A person is also embedded in a world of moral relations to other persons, is held responsible for his or her actions and in the default condition is morally protected.

Psychology is or ought to be a 'moral science' – that is concerned with research into implicit and explicit norms and rules for the conduct of life – moral (*taking care of the sick*); prudential (*exercising in moderation*); and practical (*remembering not to add water to hot fat*). In using positioning theory to disentangle the fine structure of local orders, there are many contexts, such as farming or playing tennis, in which personal embodiment is a relevant consideration, but by no means in all.

To defend the idea of this type of psychology we need to propose and defend the utility of some metaphors with which to shape research programmes where personal embodiment is the relevant feature. One such metaphor is that of the 'site for a person' as distinct

from 'person'. For example, Michael Schiavo defended his decision
to terminate the life support for his 'brain dead' wife, Terri Schiavo,
despite the protests of her family and the disapproval of the President
of the United States, by claiming that the body in the hospital bed was
uninhabited by a person, and only persons are fully morally protected
(Grattan 2003). Terri Schiavo's body was a necessary feature of the
original establishment of a person Vygotsky-wise, but the skills
and abilities that were distinctive of personhood had ceased to be
displayed at that site. Using the body-as-site-for-person metaphor,
the person had disappeared. Perhaps the relation 'site of' captures
this melancholy history somewhat better than any other. A site and
the building erected on it are strongly related, for example they
have the same street address, but the site is not part of the building.
Demolition of the building can occur without destroying the site.

Another metaphor for managing research programmes in
cultural-discursive psychology in which embodiment is an important
consideration is that of 'person as agent undertaking a task' and the
body and its parts as a tool kit for accomplishing some of them.
The utility of the task-tool metaphor should be tested in the context
of the problem of devising research programmes in which neuroscience and cultural-discursive psychology are both used to explore
the nature of mental disturbances, as defined by the current norms
of society that appear in the activities of people going about their
everyday lives. Here we have the important insight that social order is
most visible when we are confronted with disorder. An elderly person
no longer recognizes the members of his/her family. A man claims
to hear voices urging him to carry out socially and morally unaccep-
table actions. The psychiatrist proposes the hypothesis that there is
something abnormal about the brains of such people. An axe bounces
off the log to be split. The woodman proposes that the blade is in need
of sharpening. Youthful remembering, acceptable modes of action
and easy firewood splitting are dependent on the right working of

the equipment, where 'rightness' is determined by the propriety or standards of the activity being performed in the relevant form of life.

The delineating of a cognitive task such as remembering one's family, and the standards that must be met for it to be well done in a certain cultural-historical context, are independent of neuroscience. While failures to perform 'correctly' are sometimes due to ignorance or misunderstanding or wilful rejection of the local normative framework, they are sometimes due to a defect in the 'machinery', some aspect or part of the person's body that we can consider the tool or one of the tools by means of which the proper task is to be carried out. Adopting this metaphor is a way of shaping hybrid research projects, such as those that have revealed the role of the chemical serotonin re-uptake inhibitor in the management of the socially and the culturally defined form that depression currently takes.

However, if we are to follow up the metaphor of the brain and its organs as tools for carrying out culturally defined tasks, we must deal with the objection that while we attend to the lawnmower while mowing the lawn, we don't attend to the auditory cortex while playing the clarinet. While we are attending to what we are writing, say as a calligrapher, we are not attending to the pen, just using it in a skilled way. We could say the same of the auditory cortex. While attending to the task of staying in tune we are not attending to the goings on in the cochlea, hair cells and so on, but using that piece of equipment to manage our performance. We create social order as we go along, pausing for reflection only in planning the future and dealing with present moments of disorder.

A further boost for the task-tool metaphor comes from the way we use apps on mobile phones. One could use one's arithmetical skills directly, activating a brain region somewhere in the frontal lobes, or we can perform the task by switching on the phone and activating the appropriate app. There is little conceptual difficulty in taking the phone/app combination to be a tool, so by parity of reasoning there should be no difficulty in taking the brain region as a tool. In neither case need we attend to the workings of the tool, only to its overt use.

Conclusions

While the general structure of those social microsystems, like families, that make the emergence and maintenance of social order possible is ubiquitous, the details of such systems are extraordinarily varied. This variation is due partly to the exigencies of acquiring the local language and other symbolic systems, and the skills to make a life in the local environment. Compare growing up in a family of camel herders in Tajikistan with acquiring the necessary life skills in the family of an international banker.

Positioning theory has this character of a basic universal structure, in the root concepts of 'having a right' and 'having a duty', with the means for researching into the crucial differences that mark out distinct social groups. We learn what positions there are in our circle, and we begin to grasp the interactional character of the reciprocity of rights and duties and its limits. The core concepts of positioning theory come out of vernacular moral philosophy while the human family unit and its extensions into the tribe are shaped by the limitation of the possibilities of personal interaction revealed by such studies as those of Dunbar (2007).

While there are at least some universal concepts with which we think and act in social life, there are no universal laws of social order. Just think of the wide variety of devices humanity has invented for managing the relations between the sexes, and how they have changed even in the last few decades. In the last analysis, the world of human social order is a world of persons, as embodied, conscious symbol using, and morally protected and accountable individuals.

References

Bennett, M. R., and Hacker, P. M. S. (2003), *Philosophical Foundations of Neuroscience* (Oxford: Blackwell).

Bruner, J. S. (1986), *Actual Minds, Possible Worlds* (Cambridge, MA: Harvard University Press).

Comte, A., (1909 [1861]), *A General View of Positivism*, trans J. H. Bridges (London: Routledge).

Dunbar, R. (2007), *Evolution and the Human Sciences* (London: Sage).

Freud, S. (2002 [1929]), *Civilisation and its Discontents* (London: Penguin).

Grattan, K. (2003), 'The Dispute over the Fate of Terri Sciavo ', in R. Harré and F. M. Moghaddam, *The Self and Others* (Westport, CT: Praeger), 113–45.

Harré, R., and van Langenhove, L. (1999), *Positioning Theory* (Oxford: Blackwell).

Harré, R., and Rossetti, M. (2011), 'The Right to Stand the Right to Rule', in F. M. Moghaddam and R. Harré, *Global Conflict* (New York: Springer).

Hegel, G. W. F. (1967), [1821] *Hegel's Philosophy of Right*, trans T. M. Knox (Oxford: Oxford University Press).

Hobbes, T. (2012 [1651]), *Leviathan*, ed. N. Malcolm (Oxford: Oxford University Press).

Locke, J. (1698), *Two Treatises of Government* (London: Amersham & Churchill).

Mill, J. S. (2003 [1859]), *On Liberty*, ed. G. Katcb (New Haven, CT: Yale University Press).

Smith, A. (2008 [1776]), *The Wealth of Nations* (Radford, VA: Wilder).

Spencer, H. (1900 [1862]), *System of Synthetic Philosophy: First Principles* (London: Williams and Norgate).

Vygotsky, L. S. (1975), *Thought and Language* (Cambridge, MA: MIT Press).

Part Three

Free Will after the Laws of Nature

7

Freedom and the Causal Order

T. J. Mawson

Editorial Link: The discussion of forms of distinctively social explanation naturally leads on to the consideration of one of the most
distinctive claims made about human beings, that they are free and
responsible. On a classical 'law of nature' account it is very difficult
to reconcile libertarian free will with the universal rule of laws of
nature. Philosopher Tim Mawson defends the view that there exists a
form of irreducible top-down causation. That is, the complex physical
constitution of human persons generates a new capacity of such
conscious organisms to be real causes, in addition to all the physical
causes within the system. For him, there do not have to be additional
immaterial souls for this to be possible. But laws must be seen as
non-determining or 'open', and there must be additional forms of
'top-down' causality which cannot be reduced to lower-level physical
causes. Both these features entail a rejection of the reductionist and
purely physicalist view of laws of nature, though they introduce no
'supernatural' or purely non-physical causes either. KW

You are an astronaut; the year, 3014 AD. You are exploring many
light-years past any recorded human settlement when your one-man
spaceship undergoes catastrophic failure. Just before it explodes, you
manage to crash-land it on a planet and stumble away. Now your luck
changes for the better. On stumbling from the wreckage, you discover
that the planet not only has a breathable atmosphere, it is generally
Earth-like and habitable. And, not only is it habitable, but it is in
fact already inhabited by a friendly and apparently thriving society

of humans. They descended from the crew of another spaceship that crashed several decades previously. It is in their society – which they call 'The People's Republic of Freedom' – that you will need to live out the rest of your years for you soon discover that, despite this society's advanced technology, there's no way for you to communicate your amazing survival back to Earth or indeed for anyone from Earth ever to discover you. So, what's this society like? Very pleasant.

In The People's Republic of Freedom, the citizens happily share the duties of agriculture; working to maintain and improve infra-structure and the environment; raising families; and caring for the sick and elderly. Their own good efforts and the technology available to them mean that they have ample time to pursue without restriction whatever religious, artistic and scientific projects they wish. There is never any crime; there is never even any disagreement.

You are guided around this utopia by its avuncular creator, who tells you of these features of it whilst he genially presses you to drink glasses of the country's favourite beverage, Freedom Froth. After a while, he considerately asks you if you'd like to sit down. The Freedom Froth is indeed making you rather light-headed, so you do so. You sit on a bench that is next to one of the society's ubiquitous 'co-ordinating computers', as the Creator calls them. As you sit there, the Creator tells you about the history of the society of which you will from now on be a member. This is what he says:

> I was always impressed by Mill's ideal that the state should try to grant to each citizen the maximum freedom compatible with a similar level being held by every other. But, at the same time, I was concerned that meeting that ideal by itself would not prevent there being an upper bound on the amount of freedom that each citizen could enjoy, an upper bound generated by the fact that citizens might have conflicting desires or make, as we might say, 'conflicting choices'. You might be familiar with the thought as expressed casually with words such as, 'Your freedom to extend your arm must finish just prior to your fist

hitting my nose'. The presumption of such a case of course is that a citizen who chooses to extend his or her arm might find himself or herself in close proximity to a citizen who has the desire not to be hit on the nose. It thus quickly occurred to me that society could only be maximally free – this upper bound could only be removed – by eliminating conflicting desires and choices. But that, it struck me, was no physical impossibility; it was just a neurological engineering problem and I happened to have a suitable background. After our ship crash-landed on this planet – in the first few years, whilst the others built shelters and farms – I thus spent my time working on two projects that, in conjunction, have enabled me now to create a society of people about whose freedom no worries can legitimately be raised.

Despite now feeling rather woozy, you lean forward to make sure you hear all that the Creator goes on to say; from somewhere deep within you, a sense of unease is struggling up towards your consciousness. The Creator continues:

The first project was a series of computers of such sophistication as to be infallible about what desires being had and choices being made by what citizens at what time would eliminate conflict and best enable the society to continue on in existence. I combined these with transmitters capable of beaming this information in a targeted way into the heads of the relevant citizens. The co-ordinating computers you see around you are the fruits of this. The second project I'm particularly pleased with, as I was able to make it into a pleasant-tasting beverage. It's a drug which attunes people's brains to pick up on this information and necessitates that, from within five minutes of their first drinking it, they can only ever have the desires the computer legislates and make the choices the computer decrees for them. Finally, I completed these projects and thus The People's Republic of Freedom was born. It is a society in which every citizen has maximal freedom – he or she can literally do whatever he or she wants or chooses – as a result of his or her being incapable of wanting or choosing anything other than whatever it is the computer tells him or her to want and choose. Your scanning by the computer next to which you sit is complete.

Your first instructions are already being transmitted. The five minutes needed for the drug you have drunk to take effect are almost up. Very soon now, you will mesh in perfectly with the rest of us, being incapable of wanting, choosing, or indeed – joy of joy! – thinking or believing anything other than what the computer determines you to want, choose, think and believe. You will soon then being incapable of having your freedom to act on your wants and beliefs frustrated by anyone or anything else.

The feeling of unease that was growing in you is now taking the shape of a more determinate thought even as you become aware that the drug is making you care less about it. With a last effort, you try to articulate it. The words are almost there, but they seem somehow stuck on your lips. The creator pauses, noticing the look on your face. He asks you a question. 'What are you worried about?'

What you're worried about in this science-fiction scenario is losing the sort of free will that this chapter is about.

So far, so much science-fiction. What about science fact? It's now commonplace to observe that there has been a shift in scientific consensus over the last few hundred years on an issue that seems relevant to whether or not we have free will of the sort that it seemed you were about to lose in my science-fiction scenario. Classical, Newtonian mechanics, which lent itself most naturally to a deterministic interpretation, ruled scientific thinking for centuries. But quantum physics, the most popular ('Copenhagen') interpretation of which is indeterministic, has now displaced it. We may still use classical mechanics as a good approximation for everyday purposes; you can even land a spaceship on the Moon with it. But the consensus now is that technically – and in every sense at the sub-microscopic, quantum level – it's not a true account of the world.

A common layman's reaction to this changed consensus is that it somehow makes it easier than it used to be to maintain that we have free will of the sort this chapter is about. When the world was seen as Newton saw it, there was no natural place for us as agents to make a difference; now there is: at the sub-microscopic, quantum level. Of course it was possible, even in the heyday of Determinism, to posit that we had souls – supernatural substances which miraculously intervened, violating these laws of nature so as to produce our actions. But such a posit seemed the worst sort of panicky metaphysics; it seemed to fly in the face of all that the science of its day was telling us. By contrast, now that we may see ourselves as living in an indeterministic universe, that, somehow, means that we've no need to posit supernatural substances or miracles in order to maintain that we have free will. And thus our understanding of ourselves as free in the sense we're thinking about now fits in with, rather than clashes with, what science is telling us. This is certainly a common reaction and, to cut a long story short, I think that it's the right reaction. But the 'somehows' in it need investigation. For it is not obvious quite *how* our universe being indeterministic can really help us be free in the relevant sense. Indeed Indeterminism seems to bring its own problem – the danger of 'mere randomness', as we might put it.

The only way for Indeterminism to be true is for some of the events that happen in the universe *not* to have been causally necessitated by what precedes them; that's just what Indeterminism is. And the only way for this to happen is, it's seemed to some, for there to be a certain amount of randomness, introducing 'noise', if you will, to the system. But noise isn't going to help us be free any more than causal necessitation. For an agent to be free in doing an action, it has to be the case that it's *the agent* causing the action, not that nothing is causing it. And this is why mere appeal to the truth of Indeterminism is not enough. We need not simply to deny that the universe is ordered by the sorts of all-encompassing laws of nature that figured in the Newtonian world view. We need also to give an account of the sorts of order into which we as agents fit and show how it is that in an indeterministic universe there can be a sort of order that doesn't collapse into mere randomness. Only then can we rest content with saying that the layman's reaction is right: given the move to an indeterministic world view, free will – as we most want and believe ourselves to have it – is indeed alive and well.

Let's stand back for a moment and reflect at the most abstract level on what sorts of things can be causes.

Needless to say, there are a variety of views on the nature of causation and on the sorts of things that can and cannot be causes. What I shall be saying will be controversial. Most would agree that abstract objects – such as the number seven – cannot cause anything. Indeed causal effeteness is sometimes made definitional of what it is to be an abstract object. And most would also agree that events – such as the firing of a gun in a particular place and at a particular time – can cause things. I've argued at greater length elsewhere (Mawson 2011) for what I say is the commonsense view with regard to agents: agents, as well as events, can cause things. This view is sometimes called the agent-causation view. Let me sketch it briefly now before going on to discuss how we may use it and the new ways of configuring the orderliness of the natural world to avoid the 'mere randomness' problem that appealing solely to the truth of Indeterminism is likely to raise. And let me start by considering an example to ground the account in a particular time and place.

Let's suppose that the event that we're interested in the cause(s) of is the First World War. On reflection, we probably want to say that lots of things caused this – a constellation of political, economic and military factors. But one thing that would certainly be high up our list of causes is an action, the action of the assassin of Archduke Franz Ferdinand in shooting him. So, if listing the full causes of the First World War – an impossible task, of course – one thing that we would wish to put on the list would be the particular agent who was the assassin; his name, in case you've forgotten it, was Gavrilo Princip.

The agent-causation view is that the events that led up to Princip' assassinating Ferdinand didn't determine in the sense of necessitating that he assassinate Ferdinand. The socio-economic-politico-military context might have been as it was; Princip could have had all the thoughts that he did; and yet *still* he might have not gone through with the assassination, deciding at the last minute not to act on the reasons that he believed himself to have. (As I recall it, another member of Princip's band had backed out of throwing his bomb at the last moment earlier in the day. Princip could have similarly backed out.) As it was though, Princip himself added to these events at the moment of his choice some causal 'oomph' if you will. And so it was that he pulled the trigger and assassinated Ferdinand, thereby starting the First World War. On this account, agents outstrip the causal powers of the parts that make them up; they produce events directly, events which are not in fact solely caused by preceding events, events such as their thinking they have good reason to do what it is they're about to do. And the shift to Indeterminism makes this account tenable. As I say, this seems to me the commonsensical thing to think about actions. But on this account, it's not that Indeterminism just introduces the possibility of randomness or 'noise' into what would otherwise have been a deterministic universe; it introduces the possibility for us as agents to be the things making the difference. Once all the non-agential facts have spoken – the events have caused the various possible outcomes to have whatever probabilities they have – there's still something or, rather, someone yet to speak – the agent. The agent can come in at this stage and make happen that which was quite likely to happen anyway, but also – though presumably less frequently – make happen that which was quite unlikely to happen. And thus we as agents enter the world of events. Okay, one might say, the move to Indeterminism allows that we as agents *could* fit into the causal order – it needn't be mere randomness – but how exactly *do* we fit into it?

If we believe in the agent-causal view, we are committed to believing that there is a certain type of thing – agents – the causal potency of which cannot be reduced to the causal potency of the events that this type of thing undergoes. What sort of thing is this? As mentioned previously, a traditional answer would have been that it is a non-physical substance, a soul. But new developments in our understanding of the causal order – what we may call the rejection of reductionism – have made it possible to maintain another answer: we human beings are entirely physical substances, but we're the sorts of substance that have causal powers which are independent of the powers of the things out of which we are constituted.

Two sorts of reductionism are no longer universally accepted amongst scientists (they were never close to being universally accepted amongst philosophers). Indeed one is now commonly rejected. The sort of reductionism that it is now almost a commonplace to reject might be called 'vertical reductionism'. In rejecting this sort of reductionism, one commits oneself to the view that, in some cases, what exists at what is in some sense a more basic level of reality does not determine what happens at what is in that sense a higher level. New phenomena, indeed we may say 'substances', emerge, substances with genuine causal powers. The sort of reductionism that is still widely, but not now universally, accepted might be called 'horizontal reductionism'. In rejecting this sort of reductionism, one commits oneself to the view that in some cases natural laws are non-universal at a given level of reality. The relevant laws may dictate that such-and-such happen in one type of system at a given level, but say nothing at all about what must or can happen in another kind of system at the same level. It's not that the law applies but that miracles (understood as breakings of that law) are to be expected in one system but not in the other; it's that the law just doesn't apply in that second system, so is not 'there' to be broken any more than it's 'there' to be kept, just as the laws of Whist aren't 'there' to be broken or kept when playing Poker. It's the laws not being there that means that there's no need to break them; and it's there being no need to break them that means that we no longer need to posit a supernatural substance – a soul – to do the breaking. We might be entirely and non-problematically natural things and yet perform the same sort of role that traditionally it has been thought only an immaterial soul could fulfil.

To give a flavour of the sort of thinking going on when contemporary philosophers of science reject reductionism, let's imagine for a moment a particular pointillist painting. Obviously, at some level, the painting consists merely of a series of patches of colour – as of course do all paintings, but the pointillist style makes the fact more obvious than some. These patches of colour have various causal powers, e.g. the power to reflect light of certain wavelengths and absorb light of other wavelengths. The picture that emerges if one stands back and looks at the canvas from a distance has other causal powers: for example, it may have the power to make one realize that it depicts people relaxing on a riverbank. We may say of the higher-level entity – the picture – that it has causal powers which the lower-level entities that constitute it, the patches of colour, do not have. Similarly, then, we may suggest that humans *qua* agents are higher-level entities with causal powers that exceed the causal powers of the component molecules, or what have you, which constitute humans *qua* collections of cells.

It must be conceded that there is a natural inclination to think that higher-level entities cannot have causal powers that are genuinely independent to any extent of, or in any manner exceed, the causal powers of the lower-level entities that constitute them; there's a natural inclination to think that they couldn't have been different nor indeed have been exercised differently without a change in the causal powers of the lower-level entities or their exercise. In the case of our example, this natural inclination would lead us to think that if the painting has the power to cause the average viewer to realize that it is a depiction of people relaxing on a river-bank, then it only has it in virtue of the patches of colour having the powers to appear the colours that they do to the average viewer and, if you had wanted to paint a different painting, there'd have been no way of doing it other than by putting the dots in different places. This natural inclination is the inclination to believe in vertical reductionism. And whilst it seems perhaps right in the case of the picture, this is a natural inclination to which the believer in natural agent-causation cannot give-in when it comes to the case of agents. And here we see the rejection of the two sorts of reductionism interacting. Why would we expect the same to be true of people as is true of paintings? Psychologists don't, after all, perfect their science by practicing it on paintings.

Believers in agent causation within the physical world are committed to the actuality and irreducibility of what we may call 'top-down causation'. They are committed to the falsity of the claim that there can be no difference in a higher-level property without its being caused by a difference in lower-level property, through their being committed to the claim that the higher level is not in fact determined by the lower. We all believe in bottom-up causation: that's plausibly the story to tell about how we are caused to see a picture of people relaxing on a river-bank by the light reflected by thousands of tiny dots of colour from a painting. And we can probably be made to believe in top-down causation without much difficulty, at least as 'a useful story' to tell in everyday life. That's pretty-obviously the story to tell about how, having enjoyed looking at this particular painting, I end up buying a postcard of it from the gift shop on the way out, and thus why it is that the particles that constitute that particular postcard, rather than those that constitute some other postcard, move with me out of the gallery when I leave. But *irreducible* top-down causation and the claim that there can be a difference in a higher-level property without its being determined by a difference in a lower are still much more controversial notions. But however controversial they might be, these new developments in conceiving of how the natural world is ordered show at least that there is nothing incomprehensible or unscientific about them. Just as agent causation per se does not posit a new relation of causation, just that – in addition to events – this type of substance can cause things, so the person who posits irreducible and non-dependent top-down causation is not positing a new relation of causation either, just that the familiar everyday notion operates from the higher level down in a way which is not just a story – shorthand for lower-level entities producing epiphenomenal higher-level effects; in fact there can be changes in higher level properties which are not determined by changes in lower. The assertion that causation can go in this direction is no less comprehensible than the assertion that it cannot. There are no scientific findings that suggest it doesn't happen and our everyday experience suggests that it does. The details of quite how we exercise this power are things about which what we can say is perforce somewhat speculative, but let me venture the following thoughts.

One of the ways in which it is possible that we as agents act is by 'tweaking', if you will, the sub-microscopic. Perhaps as agents the effects we most immediately produce are ones of determining quantum 'measurement' events in ways that are compatible with the indeterministic laws; the laws always held that things such as this might happen. The failure of horizontal reductionism shows us how this might be, even without evidence of it happening in the lab. In sections of human brain tissue abstracted from the wider context of the normally-functioning human head and kept alive in a Petri dish, it might well be that the relevant laws of nature rule out such happenings. So be it. We are not talking about such circumstances and it may well be that the laws operative over sections of matter in such contexts are different from those operative over sections in the context of the normally-functioning human head. Indeed, surely we'd expect them to be so. Psychologists, after all, don't perfect their science any more from practicing it on sections of brain matter in labs than they do from practicing it on paintings. In the case of the sections of brain tissue in the normally-functioning human head, it might well be that these sub-microscopic events which we top-down cause directly then have their causal influence amplified – moving bottom-up – so that they have the significant (and primarily intended) consequences at the macroscopic level. A variant on this model would have us as agents intervening (but not violating any laws by doing so) in a relatively large number of places more or less at once at the relatively macroscopic level in the brain, the relatively macroscopic effect – e.g. the brain-state that then determines my arm to rise a moment later – being produced directly; there is no need then for any amplificatory structures. Indeed, having produced this effect at the macroscopic level, it may be that microscopic and sub-microscopic effects flow from *it*, through top-down or whole-part causation. Any of these ways could be ways in which we as agents most basically enter the world of events at particular moments. But, as well as acting at particular moments, we act over time and – importantly – we act on ourselves. Over time, it may be that certain of our choices are able to affect brain structures, brain structures from which we then act 'by habit' later. Surely it is indeed obvious that we do train ourselves at earlier times to act from habit in certain contexts and that at later times that training is shown to have worked – without thinking, we act on it. We may wish to give some such habits the honorific 'virtues' (and others the less-honorific 'vices'). Then, when we act as we might say 'from these virtues', without thinking, it can still genuinely be us who acted. But in other circumstances and more basically it will be as direct causes that we are most aware of ourselves as acting; we will be conscious of ourselves at a moment choice, unguided by habits or choices we have made previously.

So, this is the view that, as I say, seems to me to be the common-sensical one to hold about us as agents and this is a sketch of how new ways of conceiving of the natural order have made it easier to hold. Finally, I turn to consider how so seeing ourselves gives us an insight into why we were right to 'fear' determinism, i.e. to believe that 'had our universe turned out to be deterministic, we wouldn't have been able to maintain that we had free will'. To this end, it is profitable to ask whether agents are the sorts of things that themselves can feature as effects. I've argued that as well as events being causes of effects, agents can be causes of effects. Can agents be the effects of causes?

There is one – and only one – context in which it appears that they can. It seems natural to say that parents cause the agents that are their offspring to come into existence and we may express this fact without its sounding too unnatural as their 'causing agents'. Be that as it may, once an agent is in existence, the only thing that other agents and events can cause with respect to him or her is for him or her to undergo events; once an agent has started existing, he or she cannot in any way himself or herself be caused, be the effect of other substances, actions and events. This is true even though, of course, other agents and events can cause him or her to cause something else, through e.g. inclining him or her towards a particular outcome by presenting him or her with reasons to favour that outcome; and – through self-forming actions – we can cause ourselves to become more or less virtuous, for example. That being the case, once we are considering an already-existing agent, we may say of him or her that nothing can cause him or her; he or she cannot himself or herself be the effect of anything preceding him or her. And that being the case, if he or she then goes on to cause some effect, that effect *must* be undetermined by preceding events – because it was caused by him or her and he or she was undetermined by preceding events. It is this then that explains most directly why it is that we were right to fear that were our universe to have been deterministic, we wouldn't have been able to perform genuinely free actions. Genuine actions require agent-causal oomph, as I put it earlier, either and most obviously at the moment of action or earlier, when we developed structures from which we later acted, habits as I have called them. And the event that is the agent providing agent-causal oomph has as its initiator – obviously – a pre-existing agent, a pre-existing agent being the sort of thing that by its nature cannot itself be the effect of any cause. That – most fundamentally – is what *would have been* denied us *had* our universe been deterministic. As it is though, our universe *is* indeterministic; thus we *can* be the things that make the difference. Of course we can only do so within limits, but Indeterminism's being true means that these limits don't always shrink the possibilities open to us down to one; sometimes, they leave a number of possibilities greater than one open to us and then, either through some character trait, a habit or in a way that is more paradigmatic of free choice, we may act. No doubt that makes us special – it seems as if no non-animal has this ability. It is even arguably the case that no non-human animal has it. But by no means does it make us unnatural. New ways of conceiving of the natural causal order mean that we've no need to posit souls, for example, performing little miracles (although these new ways are compatible with souls performing miracles) or that the only alternative to causal necessitation is mere randomness.

This, I have suggested, is the sort of freedom which we enjoy and which you were afraid of being about to lose in the science-fiction scenario that I started the chapter by asking you to imagine. In addition, it is a sort of freedom about which your imagined fear in that thought experiment reveals something else. Your imagined fear at the prospect of losing it reveals that this sort of freedom is something that you think that it's good that you have.

References

Kane, R. (2005), *A Contemporary Introduction to Free Will* (Oxford: Oxford University Press).

Mawson, T. J. (2011), *Free Will: A Guide for the Perplexed* (New York: Continuum).

8

From Laws to Powers

Steven Horst[1]

Editorial Link: Steve Horst complements Mawson's argument by a fuller consideration of the nature of laws of nature. He too wishes to defend the possibility of libertarian (non-determined by prior causes) free will along with a transformed view of the laws of nature. But whereas Mawson gives some importance to quantum indeterminacies or probabilistic laws of nature, Horst concentrates on non-probabilistic laws like gravity. He argues for a view he calls 'cognitive pluralism' – laws of nature provide 'potential partial causal contributions to real-world kinematics'. Even ordinary non-probabilistic laws of nature do not just mirror the real world. They abstract and idealize, providing 'idealized claims about the actual world', approximations that only fit exactly in artificially isolated contexts. They are 'ways of representing the world', not 'features of the world'. They ignore all the many other influences of different sorts that may exercise causal influence in the real world. This is similar to what Stephen Hawking calls 'model-dependent realism' in his book, *The Grand Design* (2010). Horst, like Cartwright, thinks that it may be better to talk of causal powers, dispositions or propensities than universal rules. For such capacities would only be realized in specific contexts, and we could never say when there were other powers not covered by such laws (including, perhaps, free acts of will), or how such capacities might combine in differing contexts. Thus to think of the real world as bound by exceptionless laws is actually false – a possible view, perhaps, but not one supported by the actual practice of the sciences when they are dealing successfully with the real world, and not one that is theoretically very appealing. In particular, human consciousness and freedom call for a much more comprehensive account of order in nature which can include such factors as an integral part of the natural world, not as an inexplicable afterthought or even an illusion. KW

Is a commitment to scientific laws compatible with a commitment to libertarian free will? Since early Modernity, there has been a substantial movement towards assuming that the answer is 'no', in spite of the fact that some of the most influential early proponents of the idea that there are laws of nature (such as Descartes and Newton) were also staunch proponents of free will. I shall argue that Descartes and Newton had it right: properly understood, natural laws are completely consistent with libertarian freedom, and advocates of free will have nothing to fear from scientific laws. The *illusion* of incompatibility, I shall argue, is due to the influence of a particular philosophical account of the nature of scientific laws – the empiricist account. But the empiricist account of laws is neither the *only* nor the *best* account available. Indeed, I shall argue that it is utterly untenable, and that alternative accounts that claim that laws express causal powers, dispositions or propensities are clearly compatible with libertarian freedom. I shall introduce a cognitivist version of such an account as the most attractive account of laws.

Caveats

Because the philosophical literatures on free will and on natural laws are already truly immense, it is important to stress at the beginning just what issues will be considered in this chapter, and which will not.

1 Of the many philosophical notions of 'freedom', I am concerned specifically with the notion sometimes called libertarian freedom.
2 I shall assume (what I take to be a mainline but not universal view) that libertarian freedom is incompatible with determinism.

3 I shall not attempt to argue that one *ought* to be committed either to libertarian freedom or to scientific laws. I shall simply argue that one can be committed to both, simultaneously, without inconsistency.

4 In particular, I shall not attempt to comment on the myriad other topics on which philosophers have written concerning free will, (in)determinism or (in)compatiblism.

5 I shall not attempt to untangle the issues surrounding quantum mechanics, indeterminism and freedom. When I speak of 'laws' here, I shall not be speaking of probabilistic laws, but of the kind sometimes characterized as 'strict' or 'deterministic' (though for reasons that will become apparent, I do not favour these characterizations of non-probabilistic laws).

6 Some scientific laws are concerned with relations between variables at a single moment (static laws). I shall not be concerned with these, which are not directly relevant to freedom and determinism, but with those concerned with change and causation (kinematic and dynamic laws).

What do laws have to do with freedom?

At least since the eighteenth century, many philosophers and scientists have believed the proposition that the universe is governed by natural laws to be in conflict (or at least in tension) with the proposition that humans possess libertarian free will. The tension between these two claims played a memorable role in Kant's philosophy, and the theme has been repeated by more recent philosophers. A. J. Ayer locates the problem of free will squarely within the context of the assumption that our actions are governed by natural laws:

When I am said to have done something of my own free will it is
implied that I could have acted otherwise; and it is only when it
is believed that I could have acted otherwise that I am held to be
morally responsible for what I have done. For a man is not thought
to be morally responsible for an action that it was not in his power to
avoid. But if human behaviour is entirely governed by causal laws, it
is not clear how any action that is done could ever have been avoided.
It may be said of the agent that he would have acted otherwise if the
causes of his action had been different, but they being what they
were, it seems to follow that he was bound to act as he did. Now it
is commonly assumed both that men are capable of acting freely, in
the sense that is required to make them morally responsible, and that
human behaviour is entirely governed by causal laws: and it is the
apparent conflict between these two assumptions that gives rise to the
philosophical problem of the freedom of the will.

According to this formulation, human actions are determined because
they fall under the scope of natural laws. And if they are determined,
they are not free.

Timothy O'Connor likewise characterizes determinism in terms of
natural laws, as 'the thesis that there are comprehensive natural laws
that entail that there is but *one* possible path for the world's evolution
through time consistent with its total state (characterized by an
appropriate set of variables) at any arbitrary time.'

Peter van Inwagen gives a characterization of determinism that is
similarly framed in terms of natural laws. He defines determinism as
the thesis that,

> For every instant of time, there is a proposition that expresses the state
> of the world at that instant; if p and q are any propositions that express
> the state of the world at some instants [and q describes a state later
> than that described by p], then the conjunction of p with the laws of
> nature entails q. (gloss in square brackets added)

A few pages later, van Inwagen puts forward an argument for incompatiblism. Consider the question of whether some person J could

have raised his hand at time T. Let L be the set of laws, P a complete actual state of affairs at T that includes J *not* raising his hand at T, and P_0 be a complete actual state of affairs at some earlier time. van Inwagen then argues,

1 If determinism is true, then the conjunction of P_0 and L entails P.
2 It is not possible that J's raising his hand at T and P be true.
3 If (2) is true, then if J could have raised his hand at T, J could have rendered P false.
4 If J could have rendered P false, and if the conjunction of P_0 and L entails P, then J could have rendered the conjunction P_0 and L false.
5 If J could have rendered the conjunction of P_0 and L false, then J could have rendered L false.
6 J could not have rendered L false.

    ...

So (7) If determinism is true, J could not have raised his hand at T.

I would draw attention in particular to the fact that van Inwagen frames the argument in such a way as to highlight the ideas that, according to the determinist, (a) laws and complete state descriptions together imply unique future state descriptions for each subsequent time, and that acting in a fashion incompatible with that future state description would (*per impossibile*) 'render [the laws] false', and (b) we are incapable of rendering laws false. The view of laws suggested here is one on which *exceptions* to the laws are equivalent to *falsifications* of them.

The basic line of argument can be summed up as follows:

1 *Nomic determinism:* A commitment to scientific laws implies a commitment to determinism.
2 *Incompatiblism:* Determinism is incompatible with libertarian freedom.

Therefore,

3 A commitment to scientific laws implies a rejection of libertarian
   freedom.

As stated above, I shall accept incompatiblism for purposes of this
chapter, and hence shall concentrate on the rationale leading to
nomic determinism.

Classical mechanics was generally regarded as employing laws
that are deterministic. The use of stochastic equations in quantum
mechanics has spawned a cottage industry of discussions of (a)
whether these are best interpreted as implying that some or all of
the most basic dynamic processes are in fact non-deterministic, and
(b) whether quantum indeterminacy would provide room for free
will (Kane 1996) or perhaps actually render it incoherent (Pereboom
2001). Given the variety of interpretations of quantum mechanics
(some, but not all, of which imply objective indeterminacy), I shall
avoid these issues here. My concern, rather, is with the implications
of non-probabilistic laws, such as gravitation. My claim is that these,
properly understood, do not imply determinism. But before arguing
this claim, I wish first to turn to the question of why anyone would
ever have thought that they did so.

Nomic determinism and the empiricist account of laws

There is a very familiar account of the nature of laws on which there
is a straightforward connection between laws and determinism. This
is the account of laws made popular by the logical positivists and
logical empiricists, which I shall refer to as the 'empiricist account
(of laws)'. The core commitment of this account is that scientific laws express universally quantified claims about the real-world

behaviour of objects and events found in nature. In its original form, framed in quantified predicate calculus, this involved a commitment to the material truth of such claims. In subsequent formulations, the material claims have generally been understood to be modally strengthened to cover counterfactuals. As the modally-strengthened versions of the empiricist account entail the non-modal version, refuting the non-modal version will suffice as a refutation of the modal versions as well.

It is an unfortunate curiosity of the history of philosophy of science that the 'paradigm' examples used to elucidate the empiricist account of laws – e.g. 'all swans are white' – bear little resemblance to the laws one actually finds in the sciences. However, the account was clearly intended to apply to *real* laws, like classical gravitation, as well, even if attempts to actually spell out how such laws would look if rendered as quantified claims are the exception rather than the rule. Informally, they would be such claims as that any two bodies with mass will behave (that is, move) in a fashion described by the inverse square law. (And similarly, *n* bodies will behave in a fashion described by the versions of the gravitation law formulated for *n*-body problems. These are computationally intractable where $n > 2$, but are nonetheless 'deterministic' in the sense to be understood for Newtonian mechanics.) In other words, laws (thus interpreted) express materially true universal claims about the real-world behaviours of objects. In cases where the 'behaviours' in question are *motions*, the laws are thus interpreted as making universal claims about *kinematics*.

Interpreted as true universal claims about real-world kinematics, natural laws *would* pose problems for free will. Because of the universally quantified nature of the claims, any object to which at least one true law applies would have to always behave as the law describes it in order for the law to be materially true. This would not, in itself, entail full-scale determinism, as there might also be objects to which no laws apply. It would, in particular, be compatible with a dualism that allowed for spontaneous (that is, causally undetermined) *thoughts*. But it would not allow for free *actions*, as actions involve the body, which is composed of a very large number of particles to which physical laws apply. Even if one wished to insist on free *thoughts*, the requisite notion of freedom would be debarred from any useful role in explaining moral responsibility. In short, on the empiricist interpretation, a commitment to laws entails a rejection of any meaningful version of the libertarian thesis.

The trouble is that, interpreted in this way, laws like classical gravitation (or relativistic gravitation, for that matter) turn out to be false. Indeed, worse than that, they have no true substitution instances. A simple example should illustrate this. Take two pieces of paper with identical mass. Fold one into a tight ball and the other into a paper airplane. Drop them from the same height above the Earth. Lo and behold, they do *not* fall at the same rate. Or take the paper ball and a piece of metal of equal weight and drop them from the same height close to a magnet. Again, they do not fall at the same rate. (Indeed, if the magnet is sufficiently strong, the metal does not fall at all, but sticks to the magnet.)

Now, *if* the inverse square law made universal claims about how objects always fall, these experiments would suffice to disprove the law. But of course they do no such thing. And we all know *why* they do not: *other* forces, such as wind resistance and magnetism, also play a role in many real-world kinematic situations. And given that most objects include charged particles, this includes just about *all* the real-world situations; so if laws claimed what the empiricist account claims, they would not just need amendment with a few *ceteris paribus* clauses to account for exceptions, they would have few if any true substitution instances. You can't get much falser than that!

Nancy Cartwright (1983) has in a similar vein claimed that the laws of physics 'lie' – that is, state falsehoods. But perhaps it would be better to say that, *if the laws meant what the empiricist account claims them to mean*, they *would* be false. This, however, suggests a better way of characterizing the situation: namely, that the laws are *true* (or whatever alethetic honorific one prefers to apply to 'good' laws), but the empiricist account is mistaken about what they express.

The empiricist account of laws does entail a rejection of libertarian freedom, but the very feature that causes it to do so – that is, the fact that it treats laws as materially-true universally-quantified claims about real-world kinematics – also entails that empiricist laws would turn out to be quite radically false. There are, however, several familiar alternatives to the original, 'pure' empiricist view. These differ in the degree to which they differ from the empiricist account and in their plausibility. However, as I shall argue, each of them is either compatible with libertarian freedom, or else faces insuperable obstacles as an account of laws, or both.

Ceteris paribus laws

The most familiar adjustment to the empiricist account is one proposed by logical empiricists themselves: hedging quantified claims within *ceteris paribus* clauses. While this adjustment has often been applied only to the laws of the special sciences and in contrast with the purportedly 'strict' laws of fundamental physics, it is nonetheless available as a strategy for handling the problems encountered above. This proposal, however, has two problems: it does not yield a satisfying account of laws, and (unlike the strict empiricist account) does not entail determinism or preclude freedom.

The basic form of *ceteris paribus* laws is this: instead of simply asserting a law as a universal claim *L*, one embeds it in a *ceteris paribus* construction: 'other things being equal, *L*'. The phrase 'other things being equal' is a kind of stand-in for an unspecified list of conditions that separate the cases in which the embedded claim holds good from those in which it does not. Unfortunately, with the kinds of cases we have been discussing, 'other things' are seldom, if ever, 'equal'. The reason the quantified interpretation of the gravitation law turned out to be false was that other forces independent of gravity also make a contribution to real-world kinematics. And this is not just a fact about some exceptional cases at the margins. It is true of any interaction of bodies that possess both mass and charge, or any body which is subject to aerodynamic forces, or any of a long and open-ended list of additional causal factors. Interpreted in this way, most physical laws would turn out either to be completely vacuous or at best to apply to rare and specialized situations. And, perhaps worse, this would make the scope of the laws quite different from the scope accorded them by the scientist. Interpreted as a universal claim embedded in a *ceteris paribus* clause that excluded cases where non-gravitational forces were at work, the gravitation law would say nothing at all about cases in which additional forces contributed to real-world kinematics. But in fact the gravitation law *does* say something – and says something *true* – about cases where there are multiple forces at work. (Just *what* it says will need to be developed below.)

Moreover, once one introduces *ceteris paribus* clauses in this way, laws no longer entail determinism or even the denial of free action. The content of *ceteris paribus* clauses is generally left indeterminate. (And for good reason, as there may be yet-unknown situations in which a generalization fails to hold good.) But this opens the possibility that, among the factors that might make 'other things unequal', one might find cases of anomic causation in the form of libertarian free will.

Laws and ideal worlds

A second familiar variant upon the empiricist account is the claim that laws do not make true universal claims about the behaviour of objects in the real world, but rather make true universal claims about some 'ideal world' (Suppe 1989; Horgan and Tienson 1996; Giere 1999). And perhaps, if this 'ideal world' is sufficiently 'nearby' the actual world, the laws may be good enough to licence predictions that are 'approximately true' of the actual world.

This suggestion strikes me as wrong-headed for two reasons. First, it does not capture what the scientist understands laws to provide. The gravitation law does not merely make claims about some possible but non-actual world. It says things – *true* things – about the actual world as well. For example, it makes true claims about the gravitational *force* between two bodies as a function of their masses and the distance between them.

Second, the 'ideal world' in question would have to be one in which the kinematics of particles with mass is a function *only* of mass and distance (and perhaps momentum). But this would be a world with a very different ontological inventory than that of the actual world, as such a world would not contain any particles susceptible to electromagnetic, strong or weak forces. The claims of such a law would be about objects that are of little scientific interest, and it would say nothing about the majority of objects to which the gravitation law actually applies.

Moreover, interpreted as claims about ideal worlds, laws would imply nothing at all on the question of whether the actual world is deterministic. This is because such laws say nothing at all – or at least nothing exact – about the actual world. It is just too easy to stipulate conditions for a possible world that will turn out to be deterministic because one has stripped away so many features of the actual world.

One cannot then look at the possible world in question and reason from the fact that *it* is deterministic that the actual world is deterministic as well.

Still, the idea of treating laws as in some sense 'ideal' has something to be said for it. In what follows, I shall explore the possibility that laws are not exact claims about 'ideal worlds', but ideal*ized* claims about the actual world.

Summation of forces

It might be objected that the empiricist account (or my reconstruction of it) overlooks something very important that has been a part of mechanics at least since Newton: namely, the idea of *summation of forces*. In classical mechanics, when laws were said to be 'deterministic', this did *not* mean that *each* law, taken *individually*, determined the real-world behaviour of objects. Rather, scientists understood the global summation of *all* the forces acting upon objects in a system to determine the evolving kinematics of the system. Newton himself supplied the basic machinery for expressing and computing the summation of forces through the application of vector algebra, and foresaw that subsequent generations might well discover forces in addition to gravity and mechanical force. In the majority of real-world situations, such a function is not exactly and finitely computable, and in cases of classical chaos is not even good for approximate prediction. But even classical chaos is, by *fiat*, deterministic nonetheless.

Note that this variant upon the empiricist account of laws adds something to the standard logical reformulation of laws. The true quantified claims, if they are to be claims about real-world behaviour (e.g. kinematics), must not be statements of individual laws, but of composition-of-force functions. What is 'deterministic' here is *not* the individual laws, taken singly, but something on the order of a global force equation. One reason this is interesting is that the deterministic character of the system can no longer directly be read off the logical (i.e. universally quantified) form of the laws. The laws may be 'universal' in the sense that they apply to all objects with mass, or charge, etc. But this does not itself entail that the kinematics of the system will be deterministic. Determinism is a *separate* claim from that of the universal *applicability* of laws specifying the contributions of individual forces.

Indeed, adding a summation function would likely also require us to abandon the original empiricist assumption that laws are, individually, quantified claims ranging over real-world objects and their motions. A summation of forces implies that objects do not behave as the individual force functions describe. And so the domain of laws must be reconstrued in another fashion – say, as claims about *forces*. (This is the gist of the dynamic interpretation offered below.)

There are, no doubt, possible worlds that are exactly described by laws expressing the contributions of individual forces plus deterministic composition functions. The question of whether the actual world answers to this description, however, is complicated by several factors. First, given the limitations upon computability, we are faced with uncertainty in cases where real-world behaviour deviates from our best finite calculations. These may be cases of experimental error (say, failing to adequately screen off exogenous variables). They may be the work of yet-unknown additional laws. They may be manifestations of classical chaos. Or they may be intrusions of true randomness or anomic forces, including but not limited to free will. Further experimentation may reveal the first two types of causes, but it is incapable of definitively distinguishing classical chaos from brute randomness or anomic causal factors.

A more serious problem, to my mind, stems from the fact that there are cases in which there are roadblocks to the summation of forces. Cartwright (1983) has suggested that there are many instances in which we do not know how to evaluate a summation of forces. Some of these may be mere instances of present-day ignorance, but others seem to be more principled. Mark Wilson, for example, has argued that the different models used for different situations in fluid dynamics in fact have contradictory assumptions (Wilson 2006). And indeed, our two best-supported scientific theories – relativistic gravitation and quantum mechanics – are notoriously inconsistent or incommensurable. That is, in some physical situations, the two theories yield incompatible descriptions and predictions of how objects behave. Any logical reconstruction of science that attempts to incorporate them, in their present form, will thus have the unwelcome trait of generating contradictions. And such an account would not be truly 'deterministic', as it would entail multiple incompatible outcomes.

Of course, this problem is well known to physicists, and indeed motivates a very active search for a 'unified field theory'. It is possible that such a theory will be discovered, and will remove this particular roadblock. However, our present commitment to the claim that general relativity and quantum mechanics each have laws that express truths does not depend upon the viability of the search for a unified field theory. One can *believe* that there is (or will be) a means of expressing truths about gravitational, electromagnetic, strong and weak forces that will be both consistent and deterministic, and thus rule out free will, or at least free action. But such a belief is by no means forced upon us by our acceptance of the laws we actually have. It is a kind of philosophical leap of faith, either in what will be revealed in the future, or else in something that is true but can never be known. (There may, additionally, be a *regulative principle* to the effect that we should *look for* such a unification. But it would be an error to mistake a regulative ideal for a knowledge claim.)

Dynamic claims

Another possibility for saving the empiricist account would be to retain the thesis that laws express universally quantified claims, but to treat these as claims about *forces* (that is, as *dynamic* rather than *kinematic* claims). It is, indeed, possible to view alternative accounts of laws in terms of 'causal powers(/capacities/dispositions)', offered by Cartwright (1983, 1989, 1999) and others (Harré and Madden 1975; Hacking 1983) and described below, in this fashion. (Though Cartwright herself disavows the reality of 'component forces'.)

This view might serve us well in the case of laws of fundamental physics, where it is commonplace to speak of 'forces'. However, when we try to apply the notion of 'force' to laws of the special sciences – say, to psychophysical laws – the usage seems a bit strained. This, I think, partially explains the fact that causal accounts of laws tend to employ more ecumenical terms like 'power', 'capacity' or 'disposition'.

Causal accounts

A major alternative to the empiricist account of laws that has been offered in the past several decades is the view that what laws express are 'causal powers' (alternatively 'causal capacities', 'causal dispositions'). The gravitation law expresses the capacity of two masses to attract one another. Laws of chemical reaction express the dispositions of particular types of chemical structures to react and produce new structures. Often, a capacity will be manifest only in particular situations. (For example, a reaction between two compounds may take place only in the presence of a catalyst.) Likewise, the capacity for reaction expressed by a chemical law may be prevented by other conditions. Cartwright gives a simple illustration involving an acid and a base and their known reactive tendencies with a third compound. Suppose you know what happens if you mix a given substance with an acid and know what happens when you mix it with a base. You cannot tell what will happen when you mix it with an acid *and* a base just by doing some sort of vector algebra on the models for mixing with acids and mixing with bases, because the acid and base interact with one another in a way that negates their individual abilities to react with the third substance. The fact that you *can* make use of vector algebra in mechanics is in fact a special feature of mechanics.

> When two forces in mechanics are present together, each retains its original capacity. They operate side by side, independently of one another. The resulting effect is a pure combination of the effect that each is trying to produce by itself. The law of vector addition gives precise content to this idea of pure combination. In chemistry, things are different. The acid and the base neutralize each other. Each destroys the chemical powers of the other, and the peculiar chemical effects of both are eliminated. This is not like the stationary particle, held in place by the tug of forces in opposite directions. When an acid and a base mix, their effects do not combine: neither can operate to produce any effects at all.

This example illustrates two important points. First, treating laws as expressing causal capacities does not directly licence predictions of real-world behaviour, because in the real world, the exercise of the capacity may require (or be prevented by) factors not mentioned in the law. Second, the dynamic potentialities expressed by some types of laws are not suitable for composition of forces through vector algebra. This is a point that Cartwright argues at length in various publications, and it strikes me as quite significant: we may be committed to whatever it is that various laws express, piecemeal, but whatever this 'something' is, it does not directly imply determinism. (Though neither is it incompatible with it, except insofar as some laws may themselves be indeterministic.)

Laws, taken individually, do not entail determinism because (a) the capacities expressed in a law may or may not be exercised in a given instance because of factors left unspecified by the law and (b) real-world behaviour will also be influenced by other causal capacities, including those expressed by other laws, but an individual law does not say anything about what kinds of additional influences (nomic or anomic) there might be. Laws, taken together, do not licence determinism either. On the one hand, in many cases there is not a well-defined way of framing a composition of forces. On the other hand, having any set of laws in hand still leaves us essentially agnostic about whether the world is causally closed under that set of laws, or whether there are additional causal powers, whether nomic or anomic, that have yet to be specified.

On this view, our scientific understanding of the world comes *piecemeal*, through a 'patchwork of laws', each of which expresses one causal capacity, or perhaps a small number of such capacities. Science presents us with a 'dappled world', not a unified world (Cartwright 1999). Nothing in the laws themselves specifies all of the conditions under which the capacities they express will be active, nor do the laws themselves tell us how they act in combination. Sometimes we *do* know many of the relevant conditions and interactions, but this knowledge itself comes in the form of a myriad of rules of thumb and not something like the single axiomatic system hoped for by positivists like Carnap. Moreover, the list of such conditions and interactions is essentially open-ended, as experimentation and discovery continually reveal new ways in which processes *in vivo* do not work as described by the laws.

On such a view, our scientific knowledge leaves us agnostic about what remains to be discovered. The gravitation law says nothing about what other causal capacities, nomic or anomic, might be at work in a given situation, and likewise for every other law. One may still be free to opine, as a matter of philosophical taste, that all causes are nomic, or that the behaviour of the universe is fully determined by the complete set of causal capacities and the complete state description at a time. But such a view is in no way implied by the laws themselves. Discovery of the law of gravitation in no way implied that there were not other, yet undiscovered laws, such as those of electromagnetism. And our contemporary list of laws does not imply that there are no undiscovered laws today. Nor does it imply that there are no *anomic* causal factors, including whatever sort would be required for free action. Laws do not imply that the world is causally closed under those laws, and laws do not imply a rejection of freedom *because each law speaks only to a single causal capacity.*

I view this causal account of laws as a great improvement over the original empiricist account and those variations on the account that add *ceteris paribus* clauses and composition functions or take laws to make true claims only about ideal worlds. Most prominently, it allows the laws to count as *true* as expressions of causal capacities and to contribute something to our understanding of real-world kinematics in the face of the ubiquitous failure of laws to generate exactly correct descriptions or predictions of real-world behaviour.

At the same time, I have some misgivings about Cartwright's causal account. The first misgiving is about the status that is to be accorded to talk of 'capacities', 'powers' or 'dispositions'. Part of this misgiving is semantic and part is metaphysical. The grammar of such words almost compels us to treat their referents as properties inherent in individual objects. Often, however, what laws express are regularities in how a dynamic situation involving *multiple* objects unfolds. We can say 'this compound has a disposition to interact with an acid in such-and-such a way', but such a disposition is ultimately a relational disposition. That is, interactive dispositions do not lie entirely within an individual thing that is said to possess them. The metaphysical qualm stems from a concern that Cartwright's account, if read in a realist fashion, might commit us to a fundamental ontology of capacities, powers or dispositions. It is possible that such a commitment might turn out to be an acceptable option, but my own prejudice is to treat it as a less than desirable outcome.

The other realist interpretation of the causal account would be to take talk about 'capacities' to really be talk about *forces*. This, in fact, seems like a reasonable interpretation in the case of fundamental physical laws, though Cartwright herself resists it due to a disavowal of the reality of component forces. (My intuitions on this go in the opposite direction: component forces are often real, but resultant forces are mere mathematical constructions, even though they are the better predictors of kinematics.) Some readers may find a commitment to an ontology of forces as hard to swallow as I find an ontology of dispositions. But the interpretation of capacity-talk in terms of forces really cannot serve for a general interpretation of laws. Even if one is inclined to find real forces behind the gravitation law, many laws in special sciences like psychology or economics do not seem to involve special forces of their own. We need a way of cashing out capacity-talk that applies to laws in various sciences.

Much of Cartwright's corpus, however, suggests a non-realist reading, with a tension between an endorsement of a kind of pragmatism and residual traces of a realist empiricism. In the next section, I shall develop an alternative account that shares many features with the causal account, but is based upon a cognitivist rather than a pragmatist or empiricist approach to science.

Cognitivism and idealization

Earlier we considered the proposal that laws express exact descriptions of the kinematics of 'ideal worlds'. Instead, let us consider the proposal that laws express *idealized* descriptions of the *real* world that highlight the *dynamic* principles that underlie real-world kinematics.

Any observable situation in the real world is incredibly complex, far more complex than human minds are suited to understanding in its entirety. In order to find order in complexity, the human mind needs to filter its understanding through mental models that capture some abstract features at work in real-world situations while screening others out. Scientific theories, models and laws are particularly regimented forms of such mental modelling that aim at capturing real dynamic invariants in the world. Such models are *abstract* in that they are formulated around general principles. They are also *idealized* in that the world-as-understood-through-the-model will often fail to capture features at work *in vivo*. In particular, they involve at least the following three sorts of idealization:

Bracketing Idealization: A model of, say, gravity deals only with gravity, and brackets or screens off other dynamic principles at work in the real world. Bracketing idealizations thus create a gap between the application of the model to kinematic problems and the actual evolution of kinematics in the real world. However, this is simply the price that is paid for theoretical insight into individual dynamic principles. It does not *decrease* our capacity for prediction, as the road to any more accurate understanding of complexity must first pass through the formulation of models of the individual principles at work.

Distorting Idealization: Many scientific models describe their subject matter in ways that distort it. For example, familiar models in mechanics treat extended objects as point-masses or inelastic collisions as elastic, economic models treat individuals as ideally-rational decision theorists, etc. Distorting idealizations are often cognitively necessary to gain initial traction on theoretical problems, and models that contain them often remain the most elegant solutions of particular subclasses of problems. However, more complex models formulated to handle other cases (say, turbulent as opposed to laminar flow) often end up bending theoretical terminology in directions that make them formally incompatible with one another (cf. Wilson 2006).

Finite Approximation: When working in a purely theoretical context, it is often possible to express natural constants (the speed of light, the gravitational constant) with constant letters. However, to apply those models to calculations about real-world situations, those constants must be represented with a finite degree of accuracy. Often, the real values of the constants involve infinite decimal sequences, and so any finite value that stands in is an approximation. The degree of exactitude can matter in real-world predictions, and in cases of chaos there is no finite approximation that is sufficient to ensure an accurate simulation of the real-world evolution of kinematics.

Laws are not simply algebraic equations. They are framed against a background of a theory or model. For example, the equations of classical and relativistic gravitation require a background understanding of models of the geometry of space and time characterized by Euclidean and Lorentzian metrics, respectively. This, in turn, points to another characteristic of such models: *any model must employ a particular representational system.*

Scientific understanding *starts out* as a kind of pastiche of separate models because we need to understand dynamic contributions individually to gain explanatory traction. But the fact that the models are idealized and employ proprietary representational systems can produce barriers to integrating this pastiche into a single super-theory. The ways different models are idealized may, for example, result in incompatible definitions of theoretical terms (Wilson 2006), and the different representational systems that are appropriate for different problems may be formally incommensurable. When we recognize such a situation, as in the present situation with general relativity and quantum mechanics, we look for a unified theory, but whether we can find one depends not only on how the world is, but on how the human mind is capable of representing the world. There may turn out to be the case here, or elsewhere, that we cannot formulate a single consistent theory that captures all of the explanatory and predictive power that two incommensurable models afford us individually. At any rate, actual science often proceeds without such a super-model, even though the *search* for one may be viewed as a regulative ideal in scientific practice.

The view I have described is *cognitivist*, in that it treats laws (theories, models) not simply as features of the world itself, but as ways of *representing* the world. It is not, however, an anything-goes relativism. A law (theory, model) is *apt* for particular theoretical or practical purposes to the extent that it gets real-world invariants *right*.[2] The cognitivism consists merely in the claim that 'getting it right' involves, in part, *representing* it in a fashion that is useful for theory, prediction or intervention. The view is also *pluralist*, in that (a) it highlights the fact that particular laws (theories, models) are to a large extent *independent* of one another in how they provide explanatory traction, and (b) it presents the *possibility* that a comprehensive super-theory may be beyond our grasp, not because of 'how the world is' but because of how the mind goes about understanding the world. I thus call this view *cognitive pluralism* (cf. Horst 2007, 2011).

Individually, dynamic laws express *potential partial causal contributors to real-world kinematics*. They express only *partial* contributors because their bracketing idealizations leave open the question of what additional causal contributors there might be. They express only *potential* contributors because their bracketing idealizations may also include the fact that there are necessary background conditions for those causal contributions to be made.

This cognitive pluralist account of laws has much in common with the causal account. On both accounts, each law speaks only to a single factor (or a small set of factors) that can contribute to real-world behaviour. As a result, a commitment to any particular law, or any set of laws, leaves us absolutely agnostic as to whatever other sources might produce additional causal contributions: additional

nomic causes, brute randomness or anomic causal factors. This last
class, of course, includes whatever sort of causation is needed for free
will to operate. And thus, like the causal powers account, cognitive
pluralism understands laws in a fashion that is compatible with free
will.

Assessment

In this paper, I have discussed several accounts of laws – variations
on the empiricist account that treat laws as making universal claims
about real-world behaviour, an empiricist variation that treats laws as
making universal claims about forces, the causal powers account and
my cognitive pluralist account.

1 Empiricist accounts – treat laws as making materially true
   universally-quantified claims:
   a Original version: true universal claims about real-world
      behaviour of objects.
   b *Ceteris paribus* version: true universal claims about real-world
      behaviour within specified conditions.
   c Ideal worlds version: true universal claims about the behaviour
      of objects in a non-real ideal world.
   d Dynamic version: true universal claims about forces exerted.
2 Causal capacities account.
3 Cognitive pluralist account.

The original empiricist account (1a) would seem to be incompatible
with free will, or at least with free embodied actions. But it also has
the consequence that scientific laws turn out to be quite radically
false. Indeed, it is precisely the feature of the account that creates a
problem for free action – the assumption that laws make materially
true universally-quantified claims about objects and events – that

proved to be the weak point of the account for reasons having nothing to do with free will. If the original empiricist account were correct, it *would* be the basis for an attack on free will. But an attack based on a false account of laws gives the advocate of free will no cause for concern.

The addition of *ceteris paribus* clauses (1b) and the ideal worlds version (1c) was found wanting for different reasons: namely, they did not meet the test of having the resulting laws correspond with what laws do in scientific practice. They are thus no more viable than the original empiricist account. But they also differ from the original account in the important respect that each is compatible with libertarian freedom. The ideal worlds version says *nothing* directly about real-world behaviour. It says nothing about the behaviour of real physical objects, because real physical objects have combinations of properties that could not exist in the 'ideal' worlds. Even if the ideal worlds were deterministic, that would tell us nothing about whether the real world is deterministic as well. Moreover, if we interpret each law as describing an ideal world in which other laws (and the types of objects to which they apply) are absent, we must also ask whether the move from the real world to ideal worlds leaves out other features of the real world as well. Free will and anomic causation may fail to appear in the laws, not because they do not exist in the real world, but because the laws describe ideal worlds which are *different* from the real world in that respect.

The *ceteris paribus* version of the empiricist account attempted to preserve the material truth of universally-quantified claims by restricting their scope to cases where nothing *other than* the law in question is at work. This was problematic as an account of laws as they operate in the sciences because (a) laws *do* in fact often say something true even when they do not describe the kinematics of the situation (e.g. they may express the dynamic forces at work) and (b) once one brackets all the cases in which there is only a single law at work, it is not clear that much, if anything, is left. But if one allows laws to be hedged by *ceteris paribus* clauses that exclude cases where there is another *law* at work, it is unclear why one should assume that it is *only* other laws that can give occasion for *ceteris paribus* clauses. Another way that other things might not 'be equal' is if some type of *anomic* factor, such as free will, is at work. One might reject this option on grounds of philosophical taste, but there is surely nothing *in the laws themselves* that entails that events in the world depend *only* upon lawful causes.

The simple quantified dynamic account (1d), the causal capacities account (2) and the cognitive pluralist (3) account do not exclude free will, because on those accounts, each law speaks only to its own domain, and leaves open the question of what other factors may contribute to the entire causal mix, be they nomic or anomic. These three accounts are quite similar in (a) treating each law as speaking only to the dynamic contributions in its own domain, (b) leaving open the questions of whether there are also anomic contributions to real-world kinematics, and of how the various partial causal contributors work together to produce real-world kinematics, and hence (c) leaving room for free will. However, while the quantified dynamic account is compatible with there being additional anomic contributors to real-world kinematics, the fact that they are anomic would mean that they would have to be understood in a fashion rather different from quantified dynamic claims. By contrast, free agent causation might itself be seen as a form of 'causal power', or might be understood in cognitivist terms as a way human minds conceptualize the actions of agents that is orthogonal to how we conceive of nomic causation.

In short, the widespread assumption that a commitment to scientific laws entails a commitment to determinism and a denial of libertarian freedom is an artifact of the influence of the empiricist account of laws. On the original empiricist account, such an entailment actually follows. But that account is untenable in its own right, and its major alternatives present no obstacles to free will.

Notes

1   Much of the research for this paper was undertaken with the help of a National Endowment for the Humanities Fellowship for College Teachers and Independent Scholars.

2   I prefer not to use the descriptions 'true' or 'false' for models, as models are not propositions, but rather serve to define a possibility space for propositions. 'Aptness' is my preferred felicity term for models (cf. Horst 2007, 2011).

References

Ayer, A. J. (1954), *Philosophical Essays* (London and Basingstoke: Macmillan).

Cartwright, N. (1983), *How the Laws of Physics Lie* (Oxford: Oxford University Press).

Cartwright, N. (1989), *Nature's Capacities and their Measurement* (Oxford: Clarendon).

Cartwright, N. (1999), *The Dappled World: A Study of the Boundaries of Science* (New York: Cambridge University Press).

Giere, R. (1999), *Science Without Laws* (Chicago, IL: University of Chicago Press).

Hacking, I. (1983), *Representing and Intervening* (Cambridge: Cambridge University Press).

Harré, R., and Madden, E. (1975), *Causal Powers* (Oxford: Blackwell).

Horgan, T., and Tienson, J. (1996), *Connectionism and the Philosophy of Psychology* (Cambridge, MA/London: MIT Press/Bradford Books).

Horst, S. (2007), *Beyond Reduction: Philosophy of Mind and Post-Reductionist Philosophy of Science* (New York: Oxford University Press).

Horst, S. (2011), *Laws, Mind, and Free Will* (Cambridge, MA: MIT Press).

Inwagen, P. van (1983), *An Essay on Free Will* (Oxford: Clarendon Press).

Kane, R. (1996), *The Significance of Free Will* (New York: Oxford University Press).

O'Connor, T. (2000), *Persons and Causes: The Metaphysics of Free Will* (New York: Oxford University Press).

Pereboom, D. (2001), *Living Without Free Will* (Cambridge: Cambridge University Press).

Suppe, F. (1989), *The Semantic Conception of Theories and Scientific Realism* (Urbana, IL: University of Illinois Press).

Wilson, M. (2006), *Wandering Significance* (New York: Oxford University Press).

Part Four

Divine Order

9

Order in the Relations between Religion and Science? Reflections on the NOMA Principle of Stephen J. Gould

John Hedley Brooke

Editorial Link: The argument so far has been to adopt a richer view of the natural world, which can include such things as self-organization, top-down causation, emergence, holistic influences and even libertarian freedom. It has been about the natural world and about natural order. Yet in many places, and perhaps especially in Horst's contribution, the possibility of supernatural causality – that is, causal influence by a being who is not part of the physical order, but may be the supreme cause of order in the physical world – becomes a coherent possibility. Or at least the religious (or anti-religious) beliefs of scientists may play an important part in their formulation of scientific tasks and ideals. The last three chapters in the book explore these possibilities.

John Brooke examines the NOMA thesis of Stephen J. Gould, which argues that science and religion should be kept strictly apart. With the aid of historical examples, he shows how this is a difficult thesis to sustain, that science is a socially embedded activity and that there have been various ways in which religious ideals have helped to create an enduring culture of science. KW

Different conceptions of order have featured prominently in literature on 'science and religion', from the ordered cosmos of Aristotle, sanctified by Christian theology in the late Middle Ages, to the instantiation of order in the seventeenth-century mechanical analogues for nature promoted by Descartes; from the order captured in 'laws of nature', which for both Descartes and Newton were of divine origin, to the science-based natural theologies of the eighteenth and nineteenth centuries when a sense of order was inscribed in arguments for design. Even Darwin could say in the large book of which *On the Origin of Species* (1859) was a summary, that what he meant by 'nature' was 'the laws ordained by God to govern the universe'.[1] In a year marking the centenary of Einstein's general theory of relativity, we have been reminded that the orderliness of nature could, for Einstein, be reflected in principles of theory appraisal grounded in the elegance of mathematical representation, the intelligibility of the universe indicative of an ultimate mystery. In this chapter I shall briefly refer to these various ways in which orderliness in nature has mediated between theological and scientific understanding, but my main concern will be the meta-level conception of order proposed by Stephen J. Gould when, in his book *Rocks of Ages*, he argued that examples of the interpenetration of 'science' and 'religion' had been damaging to both and that a principle of non-overlapping magisteria was required to protect the autonomy of each. I shall argue that this meta-level of order imposed on the relations between 'science' and 'religion', while attractive to many as a default position, is a concept of order that proves difficult to implement in practice. There has to be an 'after' to Gould's inadequate NOMA principle.

It should go without saying that instead of speaking about 'science and religion', we should be speaking of 'sciences' and 'religions'. The philosopher who in England first proposed the word 'scientist' also insisted that there were significant differences between the sciences in the methods they used and in the fundamental ideas that shaped them. William Whewell observed that in the life sciences, for example, the idea of final cause was still indispensable. In chemistry, the idea of electrical polarity was a fundamental presupposition. In crystallography, the idea of symmetry played a distinctive role. Historical sciences, such as geology or evolutionary biology, had to devise methods for gaining access to the past, in a way that purely descriptive sciences did not.[2]

This may seem a rather abstract point with which to begin, but it has an important consequence. At a given point in time one science might be posing a challenge to a particular religion, when another might be offering support. For example, during the seventeenth century in Europe, the telescopic magnification of stars created the theological puzzle as to why the Creator had made so many stars invisible to the human eye. By contrast, the science of microscopy opened a door on a hidden world that was in the best sense of the word 'awesome'. In the meticulous structures of insects, even fleas, there was support for the idea of a divine craftsman.[3] Human artefacts, such as a fine needle, under the microscope appeared in all their imperfection. But from snowflakes to the compound eye of the fly, the works of nature were revealed in all their beauty.[4]

To emphasize a plurality of sciences is not to overlook the fact that scientific communities in their interface with the public have tried to articulate a formalism usually called 'the scientific method'. There is now a sophisticated literature on the rhetoric and politics of such discourses on method.[5] It is not impertinent to suggest that scientists do not always conform to their own idealizations since many have stated this themselves. What needs emphasizing is not only the existence of a plurality of sciences but that their representatives have sometimes reached diametrically opposite metaphysical conclusions on the basis of their respective methodologies. For example, during the second half of the nineteenth century, physicists and geologists reached contrary conclusions on the age of the Earth – not a trivial matter when the mechanism of evolutionary transformation was at issue.[6] Similarly, in the early years of the twentieth century, astronomy and biology could be at cross-purposes on the subject of extra-terrestrial life. Whereas astronomers gleefully did their probability calculations concerning the likely number of planets at similar distances from their suns as we are from ours, the co-founder of the theory of natural selection, Alfred Russel Wallace, actually used evolutionary theory to exterminate E.T. Writing rather as Stephen J. Gould later did in his *Wonderful Life*, Wallace stressed that at each point of evolutionary divergence, so great were the contingencies that it was inconceivable that the same path leading to intelligence akin to ours could have been followed elsewhere in the universe.[7]

Rather than multiply examples, we might switch for a moment to the other plurality – to the diversification of faith traditions. Even within one and the same religion, there can be very different appraisals of the sciences. There is, for example, an extensive literature on the role of Christian dissenters in promoting a culture of science and technology in seventeenth- and eighteenth-century England.[8] Puritan values have been seen as more conducive to scientific activity than the more contemplative Catholic spiritualities.[9] But it is also clear that different world faiths have had their own distinctive attitudes towards the sciences rendering any monolithic treatment suspect. It is often said, for example, that certain Hindu traditions have been more amenable to evolutionary perspectives than the classical theism of mainstream Christianity. Through a conflation of dominion over nature with domination of nature, Christianity has itself been blamed for catalysing our environmental crises.[10] Contrasts between different cultural traditions have been drawn in many ways that impinge on the evaluation and characterization of the sciences. Joseph Needham controversially argued that the absence of the idea of a personal God, legislating for a created world, might partly explain the absence in China of the abstract and mathematically expressed laws of nature that featured in the physical science of the Christian West.[11] Such differences have often been magnified for apologetic reasons. It is not unusual to find apologists for a specific religious tradition claiming that theirs alone has no problem with science whereas conflict is the order of the day in others. But the existence of diversity means that one cannot unequivocally say whether science, in some general sense, is either friend or foe to religion. As Geoffrey Cantor has remarked, when asked about the relations between science and religion the first response has to be 'whose science and whose religion?'.[12]

In this short chapter, I want to show what can go wrong when an author advocates a general formula to order the relations between 'science' and 'religion', oblivious to the diversification of the sciences and of the world's faith traditions. In *Rocks of Ages*, Stephen J. Gould proposed his magic formula that would allow science and religion to co-exist in peace and harmony. The secret, according to Gould, lies in stressing their differences, not in regarding 'religion', as the so-called 'new atheists' like to do, as failed science. If each keeps to its respective sphere of authority – the determination of facts and explanatory theories in the case of science, and of moral values in the case of religion – a potential harmony becomes possible. Only when the magisteria of science and religion are allowed to overlap do problems arise.[13] Gould's NOMA principle (that they should not overlap) is seductive but not immune to criticism. For a start, the sciences deal in theories that have impinged on religious authority, religions deal with systems of belief as well as moral values, and the prophetic strain in several religions may generate the conviction that overly ambitious claims for cultural authority in scientific communities should be open to criticism. This complicates the picture, as does the realization that the NOMA principle is not itself value-free. It has been enunciated many times in the past as a way of defending religious beliefs against the encroachment of the sciences and (of particular concern to Gould) to defend the freedom of the scientist to work without religious interference.

The problem that I particularly wish to address is the difficulty in applying the NOMA principle consistently. The trouble is that there has been a lot of cross-traffic between scientific and religious ideas historically: some constructive, some damaging.[14] Here are a few examples. A certain freeing of astronomy from philosophy was achieved by medieval Muslim thinkers, aided by attacks on

Hellenistic philosophy that were religiously motivated.[15] The quest for a geometrical harmony in the heavens, by Copernicus and then Kepler, was inspired at least in part by the conviction that an intelligible order had been imposed on the world by a Creator. On discovering the correlation between the period of a planet's orbit and its mean distance from the sun, Kepler confessed to having been 'carried away by unutterable rapture at the divine spectacle of heavenly harmony'.[16] When Descartes a few years later articulated the principle that motion in a straight line is conserved, he deduced it, or at least claimed to do so, from the principle of divine immutability. When Newton, to whom we normally ascribe the principle of linear of inertia, pondered the question whether his law of gravitation might be universal, a theological consideration informed his reasoning:

> If there be an universal life and all space be the sensorium of a thinking being who by immediate presence perceives all things in it … the laws of motion arising from life or will may be of universal extent.[17]

When his adversary Leibniz objected to Newton's science, the theology was again up front. The possibility of a vacuum in nature was denounced on the ground that this would introduce an arbitrary element into creation, since the deity could have enriched the world by making more matter. And as for Newton's postulation of the occasional reformation of the planetary system, Leibniz would protest that such interference smacked of a second-rate deity deficient in foresight.

During the Enlightenment the Christian religion was attacked on many fronts, but it is striking how often scientific activity was still justified in religious terms. Here is the great taxonomist Linnaeus, explaining why the practice of science is a religious duty:

> If the maker has furnished this globe, like a museum, with the most admirable proofs of his wisdom and power; if this splendid theatre would be adorned in vain without a spectator; and if man the most

perfect of his works is alone capable of considering the wonderful economy of the whole; it follows that man is made for the purpose of studying the Creator's works, that he may observe in them the evident marks of divine wisdom.[18]

This argument from design to a designer permeated the culture of science in the English-speaking world for some 200 years, from the middle of the seventeenth century to the mid-nineteenth. It survived longer still in the literature of science popularization.[19] It may not always have been good theology. To deduce the goodness of God from the fact that food tastes better than it needs to in order to sustain us does not mark the height of sophistication. That was one of William Paley's arguments, whose inferences from anatomical contrivance to a Contriver are usually seen as defeated by Darwin. Interestingly, however, Darwin provides the perfect example to show how cultural predispositions can shape the content of science. In his *Descent of Man* (1871), he observed how the assimilation of natural theology had put constraints on his understanding of natural selection:

> I was not able to annul the influence of my former belief, then almost universal, that each species had been purposely created; and this led to my tacit assumption that every detail of structure, excepting rudiments, was of some special, though unrecognized, service.[20]

This is an intriguing passage because it suggests that it was while under the spell of a natural theology that Darwin gave more scope to natural selection than he later wished to do.

Such examples could be multiplied. It may be worth recalling Einstein's claim that when confronted by a new physical theory he would first ask himself whether, had he been God, he would have made the world that way.[21] An insistence on the economy, elegance and beauty of nature and its encapsulation in scientific theory has often had religious connotations. The interpenetration of a science and a religion can be very subtle. If we are to believe Einstein, the

state of mind in which great scientific discoveries are made is like that
of a religious person or a person in love.

Taking the argument further, we have to consider whether some
of the examples Gould displayed to defend the NOMA principle
might, on closer inspection, turn out to be counter-examples. In
Rocks of Ages, the Galileo affair was treated in a sensitive and rather
surprising way. It would have been easy for Gould to have adopted
the common line and to have presented the tragedy as the archetypal
example of what goes wrong when the NOMA principle is flouted.
'The usual version', he wrote, 'stands so strongly against NOMA,
and marks Pope Urban VIII as such a villain, with Galileo as such
a martyred hero, that a model of inherent warfare between the
magisteria seems inevitable.'[22] Yet, drawing on Mario Biagioli's book
Galileo Courtier, Gould offered a contrasting view, sternly rejecting
what he calls the 'cardboard and anachronistic account that views
Galileo as a modern scientist fighting the entrenched dogmatism of
a church operating entirely outside its magisterium'.[23] In the spirit of
other professional historians of science, Gould was unhappy with the
entrenched dogmatism of the warfare model. Might not subscription
to the NOMA principle, however, have saved the day? We know
that Galileo approved the neat remark that the Bible teaches us how
to go to heaven, not how the heavens go. But at the same time he
argued that knowledge of science was one of the best aids to the true
interpretation of Scripture. He even went so far as to suggest that the
reference in Joshua 10.13 to the sun standing still 'in the midst of
the heavens' was more consonant with the Copernican than with the
geocentric system.[24] Galileo, in other words, does not appear to be
practising NOMA. And it may also be that his assumption of the role
of biblical exegete added to clerical perceptions of his presumption.

I turn now to the deeper question – whether Gould himself was
able to apply his NOMA principle consistently. When writing as
a science historian, which he did with enormous flair, Gould was

often concerned to stress the importance of social context for an understanding of scientific ideas. In his book *The Mismeasure of Man* (1984), he acknowledged that scientific 'facts' may not be as innocent as they seem: 'Facts are not pure and unsullied bits of information; culture also influences what we see and how we see it.'[25] Elaborating the point, he cited with approval an observation of Gunnar Myrdal: 'Cultural influences have set up the assumptions about the mind, the body, and the universe with which we begin; pose the questions we ask; influence the facts we seek; determine the interpretation we give these facts; and direct our reaction to these interpretations and conclusions.'[26]

The problem is that religious beliefs and practices are part of many cultural matrices. Does this not, by Gould's own admission, open the door to the penetration of scientific thinking by religious preconceptions? Was there a tension perhaps between Gould the scientist, who wished to divorce the practice of science from religious surveillance, and Gould the historian of science who knew perfectly well that science is a socially embedded activity?[27]

Gould might have replied that there is no inconsistency because his historical scholarship was also informed by the NOMA principle. To a certain degree this is true. One of the more appealing features of his historical writing is that he set out to rescue some of the brightest scientists of the past from the charge that their work was religiously or metaphysically informed and *therefore* worthy only of dismissal. His account of Thomas Burnet and his *Sacred Theory of the Earth* (1684) is a case in point. Because Burnet, writing in the seventeenth century, tried to show that a universal flood had been scientifically possible at the time of Noah, he has been portrayed as a misguided pre-scientific thinker, author of a 'bizarre freak of pseudo-science'.[28] Gould came to the rescue by pointing out that Burnet's account of the history of the Earth, including Noah's flood, was based entirely on natural mechanisms and that Burnet was already warning against bringing the authority of Scripture into debates about natural phenomena. Gould was particularly impressed by the fact that Burnet did not invoke miraculous intervention for the huge volume of water required. But is this a sufficient clincher to authenticate the NOMA principle? Not really, because the shape of Burnet's science still reflected a biblical understanding of Earth history. He was writing as an apologist for Christianity who wanted to demonstrate the rationality of belief in a universal flood and the rationality of belief in a relatively recent origin for the Earth, in contrast to the eternal world of Aristotle. There was certainly no absolute separation of scientific from religious authority in Burnet. He spoke of his satisfaction that those pieces of ancient history, which had been chiefly preserved in Scripture, had been 'confirmed anew, and by another light, that of nature and philosophy'.

Gould was surely correct in protesting against the view of some of his scientific colleagues that figures such as Burnet must be dismissed out of hand because of their religious motivation. But in so doing he may have overlooked some of the levels on which religious belief was still shaping the science. The metaphor of different levels is helpful because there could be separation on some but interpenetration on others. It is often said that Francis Bacon, one of the great diplomats for empirical science, had *separated* science from religion when he had argued that explanations for natural phenomena should not refer to divine purposes but to immediate efficient causes. But the one category of explanation did not necessarily rule out the other, as Gould himself showed when discussing a later *Theory of the Earth* (1795) – that of the Scottish landowner and gentleman James Hutton. Indeed Gould took delight in showing how reasoning from final causes, from wisdom and purpose in nature, had been central to Hutton's cyclic system of Earth history. Without a mechanism for restoring soil lost to erosion the human race would not have been provided for. For Hutton there had to be processes of replenishment and he was happy to speak of wisdom in the ends of nature:

The end of nature in placing an internal fire or power of heat, and a force of irresistible expansion, in the body of this earth, is to consolidate the sediment collected at the bottom of the sea, and to form thereof a mass of permanent land above the level of the ocean, for the purpose of maintaining plants and animals.[29]

A teleology in nature underpinning restorative processes also featured in Joseph Priestley's quest to uncover the mechanism whereby the quality of air, vitiated by human and animal breathing, is rendered, in his own words, 'more salubrious'. Although he cannot be credited with a full theory of photosynthesis, Priestley recognized that vegetation is the key.[30]

Gould never argued that Aristotelian or Christian final causes should be readmitted into science, but it was in the context of discussing Hutton that he showed sympathy for a pluralism in our ways of knowing:

Although theories may be winnowed and preserved empirically, their sources are as many as peoples and times and traditions and cultures are varied. If we use the past only to create heroes for present purposes, we will never understand the richness of human thought or the plurality of ways of knowing.[31]

There is clearly a question whether strict adherence to the NOMA principle and that statement of epistemological tolerance would always be compatible.

If these historical examples reveal a complexity greater than Gould's application of the NOMA principle would allow, can the principle still be maintained as an ideal to be striven for? Certainly; but was Gould himself able to sustain the ideal through the scientific controversies in which he became embroiled? To answer this question I shall briefly examine his scientific quarrel with Simon Conway Morris. Gould's particular take on evolutionary processes became well known through his book *Wonderful Life* (1991), in which he stressed the contingency and the lack of directionality. His interpretation of the Burgess Shale, in which many of the phyla of the Cambrian explosion failed to leave descendants, cohered with his emphasis on accident in the history of life forms.[32] By contrast, Conway Morris, whose work on the Burgess Shale informed Gould's account, had second thoughts and began to reduce the number of completely extinguished lines.[33] Against Gould's celebration of contingency, Conway Morris has continued to celebrate the constraints.[34] These are manifested in the phenomenon of convergence, where similar structures, a particular kind of eye for example, have evolved several times in independent lines of development. The dispute is fascinating because Morris accused Gould of allowing his personal, atheistic beliefs to colour his interpretation. Gould retaliated with the same charge, implying that there was surely a 'personal credo' insinuating itself in Morris's potentially more theistic vision.[35] Here we have a relatively recent scientific debate in which each protagonist effectively admitted that scientific theories, especially those of their opponents, could reflect religious or other ideological preferences. The really engaging point is that, far from rebutting the accusation against him, Gould acknowledged that the palaeontological evidence was necessarily read in the light of a favoured theory – a theory 'embedded (as all ideas must be) in my own personal and social context'.[36] So my question is: can the NOMA principle survive such an admission? If not, the door becomes wide open for investigating how different sciences in different faith traditions have shaped and been shaped by those traditions.[37] A start has been made in books such as *Science and Religion around the World*, co-edited by John Brooke and Ronald Numbers; but sensitive comparative studies are badly needed.[38]

Historians of science have not been as enamoured of 'laws' of history as many scientists have of 'laws' of nature. There is, however, a certain analogy between the quest for pattern and order in historical events and in the phenomena of nature. During the last 30 years or so, the historiography of science has undergone a reformation in which the order inscribed in master-narratives has been subverted in favour of more finely detailed contextualized studies.[39] These perhaps invite comparison with the greater understanding of localized scientific contexts in which the properties and capacities of matter manifest themselves differently. I have been arguing in this essay that the imposition of order on the relations between science and religion as recommended by Stephen J. Gould is too facile to capture the complexity to be found in the historical record or even in its sanitization by secular means. We sometimes forget that from the second half of the nineteenth century, when advocates of 'conflict between science and religion' first gained a hearing, there has been a tendency to erase from the historical record examples of the various ways in which religious ideals helped to create an enduring culture of science.[40] Recent scholarship has shown how conceptions of order in the sciences have been reinforced by both sacred and secular metaphysics, a point recently emphasized by Matthew Stanley in his penetrating study of the Christian metaphysics of James Clerk Maxwell and the secular agnosticism of Thomas Henry Huxley.[41]

Notes

1   Richards, in Ruse and Richards 2009: 61.
2   Fisch and Schaffer 1991. For the process whereby a diversity of research methods was translated and reduced into an ostensibly unique and privileged 'scientific method', see Harrison 2015: 145–82.
3   Harrison 1998: 161–204.
4   Brooke and Cantor 1998: 217–19.
5   Schuster and Yeo 1986.

6   Burchfield 1975; Smith 1998: 111–22 and 172–4.

7   Wallace 1904: 326–36.

8   Wood 2004.

9   Morgan, in Livingstone, Hart and Noll 1999: 43–74.

10 For a critique of this well-known thesis of Lynn White, see Harrison 1999: 86–109.

11 Needham 1969: 299–330. For a re-valuation, see Lloyd 2004.

12 Brooke and Cantor 1998: 43–72.

13 Gould 1999.

14 Brooke 1991, 2014.

15 Ragep, in Brooke, Osler and Van der Meer 2001: 49–71.

16 Caspar 1959: 282–4.

17 Cited in Westfall 1971: 397.

18 Linnaeus 1754 (trans. J. E. Smith 1786), in Goodman 1980: 18.

19 Lightman, in Brooke, Osler and Van der Meer 2001: 343–66.

20 Darwin 1871, 1906: 92.

21 Brooke and Cantor 1998: 226–7.

22 Gould 1999: 71.

23 Gould 1999: 71–2.

24 For a sensitive account of Galileo's principles of biblical exegesis, see Carroll in Brooke and Maclean 2005: 115–44.

25 Gould 1984: 22.

26 Gould 1984: 23.

27 Gould 1984: 21.

28 Gould 1984: 17.

29 Gould 1983: 92.

30 Brooke, in Brooke and Maclean 2005: 319–36.

31 Brooke, in Brooke and Maclean 2005: 93.

32 Gould 1991: 288–9.

33 Conway Morris 1998.

34 Conway Morris 2003.

35 Conway Morris and Gould 1998/9: 48–55.

36 Conway Morris and Gould 1998/9: 55.

37 For essays on this theme, see Brooke and Ihsanoglu 2005.

38 Brooke and Numbers 2011.

39 An impressive example is enshrined in Livingstone 2014, which
    demonstrates how receptivity to Darwinian evolution among
    Presbyterians varied substantially according to geographical location.
40 For a striking exception, see Gaukroger 2006.
41 Stanley 2015.

References

Brooke, J. (2005), 'Joining Natural Philosophy to Christianity: The Case
    of Joseph Priestley', in J. Brooke and I. Maclean (eds), *Heterodoxy in
    Early Modern Science and Religion* (Oxford: Oxford University Press),
    319–36.

Brooke, J. (1991, 2014), *Science and Religion: Some Historical Perspectives*
    (Cambridge: Cambridge University Press).

Brooke, J., and Cantor, G. (1998), *Reconstructing Nature: The Engagement of
    Science and Religion* (Edinburgh: T&T Clark), 217–19.

Brooke, J., and Ihsanoglu, E. (eds) (2005), *Religious Values and the Rise of
    Science in Europe* (Istanbul: Research Centre for Islamic History, Art
    and Culture).

Brooke, J., and Numbers, R. (eds) (2011), *Science and Religion around the
    World* (New York: Oxford University Press).

Burchfield, J. D.(1975), *Lord Kelvin and the Age of the Earth* (London:
    Macmillan).

Carroll, W. (2005), 'Galileo Galilei and the Myth of Heterodoxy', in J.
    Brooke and I. Maclean (eds), *Heterodoxy in Early Modern Science and
    Religion* (Oxford: Oxford University Press), 115–44.

Caspar, M. (1959), *Kepler* (London: Abelard Schuman), 282–4.

Conway Morris, S. (1998), *The Crucible of Creation* (Oxford: Oxford
    University Press).

Conway Morris, S. (2003), *Life's Solution* (Cambridge: Cambridge
    University Press).

Conway Morris, S., and Gould, S. J. (1998/9), 'Showdown on the Burgess
    Shale', *Natural History* (December/January): 48–55.

Darwin, C., *The Descent of Man* ([1871] 1906) (London: John Murray), 92.

Fisch, M., and Schaffer, S. (eds) (1991), *William Whewell: A Composite Portrait* (Oxford: Oxford University Press).

Gaukroger, S. (2006), *The Emergence of a Scientific Culture* (Oxford: Oxford University Press).

Gould, S. J. (1983), 'Hutton's Purpose', in S. J. Gould, *Hen's Teeth and Horse's Toes* (New York and London: Norton), 79–93, 92.

Gould, S. J. (1984), *The Mismeasure of Man* (Harmondsworth and New York: Pelican Books), 22.

Gould, S. J. (1991), *Wonderful Life* (Harmondsworth and New York: Penguin Books), 288–9.

Gould, S. J. (1999), *Rocks of Ages: Science & Religion in the Fullness of Life* (New York: Random House).

Harrison, P. (1998), *The Bible, Protestantism and the Rise of Natural Science* (Cambridge: Cambridge University Press), 161–204.

Harrison, P. (1999), 'Subduing the Earth: Genesis 1, Early Modern Science, and the Exploitation of Nature', *The Journal of Religion* 79: 86–109.

Harrison, P. (2015), *The Territories of Science and Religion* (Chicago: University of Chicago Press), 145–82.

Lightman, B. (2001), 'Victorian Sciences and Religions: Discordant Harmonies', in J. Brooke, M. Osler and J. Van der Meer (eds), *Science in Theistic Contexts* (Chicago: University of Chicago Press), 343–66.

Linnaeus, C. (1980 [1754, 1786]), *Reflections on the Study of Nature*, trans. J. E. Smith, in D. Goodman, *Buffon's Natural History* (Milton Keynes: Open University), 18.

Livingstone, D. (2014), *Dealing with Darwin* (Baltimore: Johns Hopkins University Press).

Lloyd, G. (2004), *Ancient Worlds, Modern Reflections: Philosophical Perspectives on Greek and Chinese Science and Culture* (Oxford: Oxford University Press).

Morgan, J. (1999), 'The Puritan Thesis Revisited', in D. Livingstone, D. G. Hart and M. Noll (eds), *Evangelicals and Science* (New York and Oxford: Oxford University Press), 43–74.

Needham, J. (1969), *The Grand Titration: Science and Society in East and West* (London, Routledge), 299–330.

Ragep, J. (2001), 'Freeing Astronomy from Philosophy: An Aspect of Islamic Influence on Science', in J. Brooke, M. Osler and J. Van der Meer (eds), *Science in Theistic Contexts: Cognitive Dimensions, Osiris* 16 (Chicago: University of Chicago Press), 49–71.

Richards, R. J. (2009), 'Darwin's Theory of Natural Selection and its Moral Purpose', in M. Ruse and R. J. Richards (eds), *The Cambridge Companion to the "Origin of Species"* (Cambridge: Cambridge University Press), 47–66, 61.

Schuster, J., and Yeo, R. (eds) (1986), *The Politics and Rhetoric of Scientific Method: Historical Studies* (Boston and Dordrecht: Reidel).

Smith, C. (1998), *The Science of Energy* (London: Athlone), 111–22, 172–4.

Stanley, M. (2015), *Huxley's Church and Maxwell's Demon* (Chicago: University of Chicago Press).

Wallace, A. R. (1904), *Man's Place in the Universe*, 4th edn. (London: Chapman and Hall), 326–36.

Westfall, R. S. (1971), *Force in Newton's Physics* (London: Macdonald), 397.

Wood, P. (ed.) (2004), *Science and Dissent in England, 1688–1945* (Aldershot: Ashgate).

10

Concepts of God and the Order of Nature

Keith Ward

Editorial Link: Keith Ward provides a historical outline of various differing views of the relation of a Creator to the created universe, and shows how different concepts of the natural order and different interpretations of the idea of God correlate. He argues that the 'dappled world' view of natural order correlates with a view of God as genuinely creative in novel ways over time, and as genuinely responsive to events in an ordered, open and emergent cosmos. Thus there is a natural fit of the new view of natural order with a dynamic, creative and interactive concept of a mind-like source of the physical world. KW

This book has suggested, with evidence drawn from a number of different disciplines, a view of laws of nature as 'dappled': local, fragile, holistic, developmental, emergent and open. This view is obviously of philosophical and scientific interest. But it is also very relevant to theologians and to those interested in questions about the possible existence of a God who creates and sustains the physical universe, and about the relation of such a God to the universe. Some scientists dismiss such questions as irrelevant, and such a dismissal has almost become part of scientific orthodoxy. But there are millions of believers in God in the world, and it is important to decide whether they are doomed to be in conflict with modern science, or whether their beliefs are intellectually compatible with well-established scientific knowledge, and may even have something to contribute to our understanding of reality.

One reason for the decline of belief in God is the rise since the sixteenth century of a view of laws of nature as absolute and all-determining. As we have seen in Part 3, this makes the existence of moral freedom problematic, and for the same sorts of reason it makes belief in some mind-like, intrinsically valuable or purposive basis of the physical cosmos, which has some causal input into the cosmos, problematic also. If our understanding of laws of nature changes, that will not constitute a 'proof' that there is moral freedom or an objectively existing God. But it may change our understanding of how mental or 'spiritual' causation may be seen as a coherent part of the causal nexus of the material world.

In this chapter I try to show how it does so, by tracing developing (or changing) ideas of God and nature in the Western tradition. I argue that acceptance of a 'dappled world' view of physical laws is actually helpful, not antagonistic or irrelevant to a specific, quite widely held, view of God. Many scientists may, quite understandably, have no interest in concepts of God. Yet many scientists, and many more non-scientists, do believe in or have a sympathetic interest in ideas of God. An account of physical laws which allows for the possibility of mental or quasi-mental causation at various levels may suggest an expanded view of physical nature which is compatible with the existence of some forms of spiritual reality, and this may suggest fruitful areas for future research.

It is possible to detect some affinities between specific ideas of God and the sorts of order that one might expect to find in the natural world. I shall begin with a roughly chronological account of the most prominent of these ideas, before indicating what sort of God is consistent with and even conducive to seeing the cosmos as a 'dappled', but not haphazard, order.

Early polytheism

The earliest recorded human beliefs in God are polytheistic, the world being seen as full of gods and spirits. It is then natural to see the world as the playground or perhaps the battleground of the gods, and since the gods are generally thought to have arisen from chaos, the natural world may be largely incomprehensible to humans, and subject to the desires and activities of spiritual beings which are almost entirely unknown. It may seem that such a world will not be highly ordered.

There are unlikely to be any absolute or universal laws of nature. Many diverse and limited powers will overlap, compete and achieve temporary compromises. There will certainly be no one ultimate principle (a Theory of Everything) from which all other laws can be derived in a consistent and coherent fashion. This will be an extreme form of pluralism about laws of nature – there may even not be any 'laws' at all, but just arbitrary decisions of the gods.

Nevertheless, many early religious systems refer to Fate or Destiny, an impersonal order to which even the gods are subject. One may perhaps find here a reference to an impersonal and necessary order behind all things, an order which may be accessed by divination or in visions, or perhaps by interpretations of the movements of the stars. But such thoughts remain undeveloped and suffused with mythical and magical elements which are alien to modern science

Aristotle and Plato

Aristotle developed a philosophical idea of God which has dominated much Western thought on the subject. For him, God was a changeless and perfect being, the best of all beings, totally absorbed in contemplation of its own perfection, but drawing physical things towards itself by 'love' or some sort of natural attraction. For both Aristotle and Plato, the ultimate reality is intelligible and value-laden rather than physical and morally neutral. It is of supreme value ('the good', which is even beyond being or actuality), and that value is an object of knowledge (it is known or appreciated as valuable). There can be only one supreme value, a value greater than any other and the source of all other values. It is not, however, a creator of everything other than itself, for matter exists as an independent and uncreated reality.

This suggests (though it does not entail) that any order in the cosmos will be limited (for we cannot know how far matter will be shaped by the ultimate form of the good), and it will be purposive (shaped by a tendency to approximate to or participate in the ultimate form, so far as is possible). Plato, who was Aristotle's teacher, thought that souls fall into matter because they are tempted by sensual desire, are reborn many times in the world and will finally return to the stars when they have passed beyond sensual desire. This requires a principle of objective moral causality, whereby good and evil choices shape the sort of life that souls live, and therefore shape to some extent the nature of the physical world in which they are embodied. Plato never tried to show how such moral causality, the existence of prime matter, the reality of the intelligible world and the possible existence of a Demiurge or designing intelligence, could be co-ordinated. And it has to be said that he was not very interested in empirical observations of the physical world, suggesting at one point that birds are reincarnations of 'empty-headed men' who think that astronomy is founded solely on empirical observations. That is not very encouraging for the idea of order in natural science.

Aristotle did not speak of reincarnation of souls, but he remained committed to the idea that physical things embody 'essential natures' which strive to realize their ideal eternally constituted forms. That is precisely the view that modern science rejected. It is, however, a sort of order, and it is an early formulation of the belief that nature is intelligible. There is a reason why things are as they are. Further, the reasons for things being as they are converge on one ultimate reason, and that reason is the intrinsic value or goodness of something existing. We might attribute to Aristotle the thought that if there is one rational God, then there are non-arbitrary reasons why events happen.

But for Aristotle those reasons will not be determining efficient causes (that is, as Dr Mawson's essay makes clear, events from the existence of which, together with immutable laws, one and only one succeeding event follows). They will be final causes, goals which entities may achieve 'more or less', which entities will tend to realize, but will not necessarily realize, or may only realize in part. This is a sort of 'organic order', of things tending to realize their potential natures.

That could be called 'purposive or axiological order', the order of things striving or tending to realize states of value. It is not very characteristic of post-sixteenth century science, though there are some indications that there could still be some life in the idea, in some (contentious) interpretations of evolution (for instance, in the much-reviled, but also revered Teilhard de Chardin). However, the concepts of value, purpose and consciousness, which are the principles of purposive order, have become problems for modern science rather than bases of explanation.

The medieval synthesis

The concepts which came to form the philosophical basis for medieval Christianity in both East and West were derived from Plato and Aristotle, yet transformed by the central idea of 'creatio ex nihilo', according to which God is the one and only Creator (cause by intention and agency) of everything other than God. This idea intensifies the problem of how evil and suffering can exist if God is good. But it also strengthens belief in the intelligibility of the universe. For medieval Christian thinkers accept the teaching of the prologue (the first chapter) of the Gospel of John that the universe is created through the *Logos,* which is identified in some sense with God. There are many possible interpretations of *Logos,* but it is clear that it is associated with wisdom, reason or thought. Thus there is one God who is supremely rational, and any universe created by that God will presumably be ordered by reason.

Since God is an intentional agent, not simply an unconscious first cause, the universe will be ordered to an end. It will have a final cause or set of final causes. And since God is supremely powerful, that end will certainly be achieved, and the universe will be an effective means to realize it. Further, since God, being perfect, was assumed, following Plato and Aristotle, to be changeless, eternal and necessarily what God is, the universe will be a 'block time' universe, issuing forth from God in one timeless act. Though we seem to move from past through present to future, in fact every time is eternally existent from God's point of view, and so the whole of time is in the moment of its creation complete from its beginning to its end.

The notion of rational principles of order in nature is clearly present in this view, and the principles will be teleological principles. They will be properly understood only when their end or purpose is understood, and that purpose may lie deep in the mind of God. There is a rational order in nature, and nature, as created by God, has purposes which may not be frustrated, and which sometimes can be known only by revelation.

The foundations of natural science, as the discovery of the rational ordering of a created world, lie here. But this ordering is seen as laid down by a conscious God who directs them to moral goals, and whose final purposes may be inscrutable. Natural science therefore still lay under the final authority of the Church's interpretation of revelation, and it is this which gave rise to disputes like that between Galileo and the Holy Inquisition (a dispute for which the Roman Catholic Church has since expressed regret). The medieval Church thus both gave birth to modern science and yet sometimes tried to prevent it from pursuing its own path to truth. That is a problem which has not even now been fully resolved to the satisfaction of all.

A Newtonian God

It was not just an accident that Isaac Newton was both a Biblical Christian and also a critic of the authority of the Church (he was, secretly, what we might call a Unitarian). After Newton's *Principia Mathematica*, the idea of law-like explanation increasingly became separated from that of purposive explanation in the natural sciences. The Aristotelian and medieval Christian idea of 'purposes *in* nature' was replaced, in Newton, by the idea of a God, different from nature, who had 'purposes *for* nature'. God, as rational and good, created states of affairs for the sake of their intrinsic value. But instead of God being the Aristotelian perfect ideal attracting things to imitate it, God now becomes, as Christians claimed, one who creates and orders all things, and so designs them to fulfil the function God has in mind and imposes upon them. A creator God is more active than a perfect but self-contained and changeless God, and if it had not been for the authority of Aristotle one might have expected a more interactive view of God, such as the one to be outlined later, to have arrived earlier on the Christian scene. Newton thought that God, as perfectly rational, would order nature in accordance with general rational principles. God would select the most simple and elegant set of laws that would produce the very complex and diverse set of effects that we see around us. And he thought of physical causes, not as states which actively bring about their effects for a purpose, but as states which (passively and blindly) fill the spaces in quasi-deductive equations laying down universal relationships between physical properties like mass, position and velocity. God lays down those laws, and ensures that entities conform to them.

The Newtonian God is not, however, bound by the laws of nature. God can break physical laws for good reason, and Newton believed that God did so. The laws hold other things being equal. Other things usually are equal, and this constancy and reliability of nature enables humans to predict, control and improve what happens in nature. This introduction of a personal creator ordering laws purposively suggests the possibility that human persons, also, might introduce personal (purposive) principles into how things go. Newton did not pursue this possibility.

A major change occasioned by this view of a rational creator God was that nature no longer had its own purposes with which humans should not interfere. God had made impersonal laws and had given humans the responsibility to understand and use them for the improvement of the human condition. Nature was no longer sacred. It was a mechanism with a purpose, and humans might be able to modify the mechanism to achieve the purpose (of happiness and human flourishing) more effectively.

But there was a tension in this view of things. If an omnipotent and wise God creates the universe, must everything not be in order as it is? How then can it be improved? Moreover, humans are part of the universe, constrained by its laws like every other entity. So how can they 'stand outside' the mechanism and modify it, since they themselves are parts of the machine?

These tensions might be resolved by a notion of development from simple elements to complex integrated structures. Intelligent consciousness might emerge naturally when physical structures become complex enough. This became possible with evolutionary theory, but reliance on such theory has two disadvantages. From the scientific side, ideas of a purposive development, of a sort of progress in evolution, are hotly disputed by evolutionary theorists. From the religious side, such ideas do not fit well with a literal interpretation of the Bible. Of course, these may turn out not to be disadvantages, if evolutionary non-directionality and Biblical literalism are both mistaken.

Newton was a child of the Protestant Reformation, and it could be argued that the Reformers' rejection of Aristotle (Luther called Aristotle 'that buffoon who has misled the Church') helped to develop further the rise of modern science. But the Reformation did not really help to resolve the tensions of a mechanistic view of nature with belief in the freedom of persons.

Both Luther and Calvin were strong predestinarians, holding that God actively causes everything to happen in accordance with the divine will. That does not entail any sort of physical determinism, since God does not have to obey laws of physical nature. But it challenges one philosophical objection to determinism (i.e. that it is incompatible with freedom), and the influential Reformed theologian Friedrich Schleiermacher was a strict determinist, even denying the reality of miraculous suspensions of the laws of nature.

Martin Luther

Since the Reformers also held that sinners are justly condemned by God, they still have the problem of how to make free moral choice consistent with wholesale predestination or physical determinism – or to put it in terms I have used, how purposive and law-like explanations can fit coherently together. Various proposals were made, but there were developments of the idea of God within Protestant theology that were to put the problem in a new light. Aristotle had viewed God as impersonal, in the sense that God was completely changeless. With the rejection of Aristotle, and with the Bible being translated and made available to the general public, more literal interpretations of Christian doctrines became prevalent within Protestant Christianity. References in the Bible to God and to God's actions were often no longer treated as symbolic metaphors, as they had often been in medieval theology. This led many Protestants to reject Aristotelian notions of a changeless God, and to posit a God who changes in response to what humans do, and who is angered or grieved by their sins, as a more literal reading of the Bible suggests.

The second generation of Lutherans developed a distinctive view of God known as 'kenotic', whereby the incarnation of God in Jesus expresses a real change in the divine nature (God sheds or sets aside omnipotence and omniscience in the person of Jesus), and shares in suffering on the cross (for traditional Catholic theologians the divine nature is incapable of such change and suffering). God becomes changeable and passible (capable of suffering).

Kenotic views of God do not entail any specific theory of the order of nature, but they do suggest the possibility of a causal order that is open to personal influences, at least partly expressing conscious mutual communications between God and created entities, rather than following an eternally predetermined order. That is characteristic of many forms of popular Evangelical Christianity, wherein God speaks, humans hear and act, and God responds in judgement and grace.

It is fairly obvious, however, that many statements about God in the Bible are not literal (that God rides on the clouds and shoots arrows at his enemies, for example), and it is also clear that, as the Creator of time and space, God is not limited by any physical form. So there is room for developing ideas of God that allow for divine change and real mutual relationships between God and the cosmos, without making God an object within space-time, or a person just outside space-time. Such a development occurred in nineteenth-century German thought, and is particularly associated with Hegel.

Evolutionary theory originated as a theological view among German proto-Idealists before it was transformed into a scientific theory by Charles Darwin.

Hegelian idealism

For Hegel, God is the absolute spirit who evolves or develops in time, expressing the divine nature in the history of the cosmos. Individual persons are aspects of this divine self-expression, as it objectifies itself in a plurality of wills, and then reconciles them into a final unity of absolute knowledge, when spirit is fully expressed and known to itself. Christian theologians have usually been wary of accepting this vastly ambitious scheme wholesale, and generally wish to distinguish God, the cosmos and finite persons more clearly than Hegel does, and with less reliance on so many abstract and obscure concepts. But ideas of the universe as something which changes God by divine knowledge of its contingent existence, of God being actively involved in the emergence of intelligent beings, of God experiencing the suffering of the world and of God actively reconciling the whole cosmos or even uniting it to the divine being, have been very influential, especially on Protestant thinking. Such ideas could be said, with perhaps a slight strain, to have some Biblical grounding in John's Gospel, which speaks of finite persons being 'included in' God (Jn 14.20 and 17.21); and they speak of God as a more active and relational being than Aristotle had imagined.

Process thought

Alfred North Whitehead, the English mathematician who was the originator of process philosophy, devised a metaphysical system almost as abstract and ambitious as that of Hegel. Whitehead presented a view of God in threefold aspect. First is God as primordial, a changeless and timeless reality containing the set of all possible worlds. But since these possibilities are not actual, God in this aspect is in a way incomplete.

God does not create the cosmos out of nothing. As in Plato and Aristotle, the world of temporal entities just always exists. It consists of billions of tiny 'actual entities', each of which has an inner reality of 'feeling' (not necessarily conscious), exists for a tiny moment, being constituted out of a set of immediately preceding entities which, as they perish, pass into the new entity, which integrates them into its own 'experience' in a new way. Then that new entity projects its new ingredient of experience onto a new set of entities, and perishes in turn. Thus the cosmos is a continual flow of momentary actual entities, always in process. There are no substances (thus 'process philosophy'), but a beginningless and endless process of ever-changing and ever-creative actual entities.

These entities are all experienced by, and indeed 'included in', God, and in this aspect God has a consequential nature, which consists of all the experiences of the cosmic entities, which are drawn from the reservoir of possibilities in the primordial nature of God, now made actual by the creative decisions of billions of actual entities, integrated in a unitary way in the divine all-including being. It can be seen that God is not the actual cause of events in the cosmos. Billions of actual entities are the causes of the stream of events which is taken up into the consequential divine nature.

There is a third aspect of God for Whitehead. God's consequential nature becomes a causal factor in the ongoing process of the cosmos. It is not a determining or efficient cause but, again rather like the Aristotelian view of final causality, a cause which 'lures' or attracts events by presenting to them a best possible future. Events are then free to accept or reject this divine influence or 'initial aim' presented by God. The process never ends or reaches a final consummation. It just goes on forever.

Open theology

Whitehead's grand metaphysical view remains in its full detail for the most part confined to a relatively small philosophical group. Yet some important features of his system have entered into mainstream Christian theology. In particular there are two factors that were virtually absent in classical Christian theology, though they might have been implicit both in the Bible and in more popular Christian thinking. They are the possibility of change in God, as God responds to new experiences and actions of finite beings, and of a conscious mutual relationship or synergy between God and events in the created order. These two factors have become abstracted from the technicalities of process philosophy, and have become important in some later, mostly Protestant, theology, especially in what is sometimes called 'open' theology, which espouses the non-deterministic nature of the God–world relation.

For such a theology, God is seen as essentially and continually creative, the originator of genuinely novel events. As omnipotent, God is not somehow the changeless actualization of every possibility (as in much traditional theology), but the one who is 'potent for' all things, who contains unlimited potentiality. Divine creativity is the ceaseless actualization of infinite possibilities, more like that of a creative artist than like a static perfection which cannot do anything other than it does (such immutability seems to be entailed by the classical simple and necessary God).

Though in Christian open theologies God is a genuine creator, God gives a degree of autonomy to creatures, so that they decide, within limits, what their future will be. God does not determine everything that happens, and perhaps does not even know exactly what will happen until it does happen. God will then respond in ways which do not destroy the autonomy of creatures, but which will ensure that the overall divine purpose in creation is not finally frustrated. God 'guides', 'influences' or 'lures' created entities, but does not unilaterally determine what they do. However, since God is the overall creator, originally choosing which possibilities in the divine mind will form a cosmos, God may not allow the process to go on forever without ever reaching any consummation. God may ensure, by what is after all a supremely powerful causality, that some divine purpose may be achieved in the cosmos, and so ensure that there is a cosmic goal which will be achieved, even though finite creative (and often evil) choices may make the path to achievement an arduous one, but may also contribute in a positive and creative way many specific features and elements of the goal which is to be achieved.

This view of God could provide some sort of explanation of how a God-created nature can be improved (or damaged) by human actions. If some degree of creativity and autonomy is possessed by creatures, it also promises to explain how general physical laws and free personal actions can cohere. Laws of nature may obtain with predictable rigour where personal creativity and free choices are not significant factors. As physical entities become increasingly integrated complex structures, creativity and choice, and also the new social reality produced by the co-existence of many diverse centres of creativity and choice, can become significant causal factors. What Dr Mawson calls 'agent causation' may be a new form of causal influence that is generated within organized complex physical structures. The agent need not be a separate non-physical entity. It could be the whole structure, exercising a 'top-down' influence on its constituent parts.

Much hard philosophical work needs to be done to make such processes clear, but it seems plausible to say that genuinely free agents are generated from complex physical structures. They are, however, not reducible to such structures. Whatever we think about the status of personal agents (traditional theology would call them 'souls'), if some process of emergence like this occurs, there must be laws to give reliability and predictability to the natural order. But they will be both open and emergent. That is, they will allow alternative possible futures, and will thus not be fully deterministic. And they will generate new properties (like knowledge and intentional purpose) whose causal properties will not be reducible (vertically reducible, as the essay by Eric Martin and Bill Bechtel puts it) to the properties of the simpler physical entities from which they have been generated.

It is intelligible to think of a God, a spiritual being of great wisdom and power, creating such a universe. There will be intelligible and coherently integrated sets of general principles, or laws, governing the physical cosmos. Those ordering principles may be necessary, in the sense that their existence is a necessary condition of the world generating the sorts of emergent intelligent beings that come to exist in it. The ordering principles will be purposively ordered, so that they will generate communities of creative and free agents capable of conscious relationship with God. That implies a general 'direction-ality' over time towards the existence of intelligent life, inherent in the basic structure of the laws of nature. Nature will be emergent and open, to allow for creative choice, by many relatively autonomous entities – including, of course, the creative and responsive choices of God, who becomes in that respect one causal factor (the main one) among others in the processes of a temporal universe. That implies a non-deterministic account of the laws of nature. And that in turn implies the presence of chance (not completely undetermined happenstance, but under-determined causal influence) in the natural order. Necessity and predictability, chance and probability, purpose and creativity, must all, in some form and to some degree, be factors in a cosmos which is created by a God of this temporally creative sort.

Each of these factors is denied by some scientists and philosophers, especially by those who would hold that all causal factors in the universe are reducible to unconscious, non-purposive, determined yet humanly unpredictable, basic forces which just happen, for no assignable reason, to be what they are. The post-Hegelian and post-process concept of God is rejected by some theologians, who think that it compromises the all-determining sovereignty of God, compromises divine omnipotence and omniscience, and blurs a crucial and total difference in kind between God and all created things.

Both of these denials may, however, be based on a preconceived view of what God or nature *must* be like. I doubt if any available set of beliefs about the ultimate nature of things, about what sort of order nature exemplifies and about what sort of God there might be, is overwhelmingly rational and plausible. There are clearly many views of what God is. However, the plurality is not just a chaotic 'anything goes' situation. Concepts of God have changed for good reason, and the number of rationally defensible alternatives is restricted.

Various key concepts relevant to this issue have changed over the centuries, and the issues are still very much open to debate. I do think, however, that if someone wishes both to believe in God (a supreme spiritual reality as the source of all being) and in the established facts of the natural sciences, a view of the natural world as ordered, open and emergent is the most plausible hypothesis.

Such a view can be said to be pluralistic, in that there may be many purposes for which the universe exists, and there does not have to be just one ultimate purpose or ultimate law. If there are purposes, however, there must be predictable means of obtaining them, so that not everything can be chance or random. But there perhaps must be chance elements (events not sufficiently determined and not selected by any finite conscious agency) by which alternative possible futures come into being, and between which intelligent agents, when they come to exist, can choose. In general, the more one emphasizes individual creativity and a plurality of intelligent agents, the more openness there will be in a cosmos, the less likely it is to be a quasi-deductive system of unbreakable laws and the more it may manifest totally original and creative characteristics.

The 'dappled world' view does not have to be associated with theism. But it is consistent with one contemporary form of theism, and if the cosmos is to be dappled rather than merely haphazard, some form of open theism could be an attractive metaphysical postulate. Independent arguments from the philosophy of science for a dappled world would provide some support for a form of open theism, and in this way science and theology would be more than what Stephen Gould called non-overlapping magisteria. They would be, as they often have been historically, inter-related in many complex ways, and sometimes, as I think in this case, they could interact to suggest a new and potentially fruitful way of seeing the cosmos.

Further reading

Most of these works exist in many editions, so specific editions and translations are not referenced.

Aristotle, *Metaphysics*, Book 12.

Hegel, W. G. F. (1807), *Phenomenology of Spirit* – VII, 'Religion'.

Newton, I. (1687), *Philosophiae Naturalis Principia Mathematica* (Scholium).

Plato, *The Republic* – particularly the Allegory of the Cave, vii. 514A–521B.

Ward, K. (2006), *Pascal's Fire* (Oxford: Oneworld Press).

Whitehead, A. N. (1978 [1929]), *Process and Reality*, Part 5, Chapter 2, 'God and the World', D. R. Griffin and D. W. Sherborne (eds) (London: Macmillan).

11

From Order to God

Russell ReManning

Editorial Link: Finally, Russell ReManning argues that seeing the universe as open to many forms of causal influence, as emergent and yet as ordered, and as creatively realizing diverse capacities that have been inherent in it from the first, can provide a positive form of argument for God as the supreme informational and ordering principle of the universe. This makes belief in God both rational and plausible in the light of modern science, though such arguments do not claim to be – as good science does not claim to be either – utterly and finally conclusive. KW

God: Out of order?

Previous chapters have traced the exciting new ideas about order in the wake of the demise of the previously dominant notion of 'laws of nature'. Different chapters have looked at new thinking about order and its implications in the natural sciences, social sciences and moral philosophy. The previous chapter by Keith Ward traced various correlations between ideas of God and of natural order. This chapter covers similar ground, but from a different perspective. I consider the prospects, given the new understandings of order, for so-called 'natural theology' – defined here as the development of arguments to prove the existence and nature of God without reliance on revelation or particular religious traditions. My claim, in a nutshell, is that the new conceptions of natural order emerging after the laws of nature offer new opportunities for natural theology; indeed, it may be possible to move from natural order to God.

This chapter has two sections. I begin by setting out a broad case for the possibility of natural theology after the laws of nature. Central to this is the recognition that whilst certain historical forms of the arguments for the existence of God are closely bound up with the kind of understanding of laws of nature developed by early-modern philosophers [such as Eric Watkins discusses in Chapter 1], this form of natural theology represents only one alternative strategy. There is nothing in principle, as Keith Ward has shown, that links the idea of God to any particular idea of natural order, that of the laws of nature included. Indeed, I will suggest that the attempt to base a viable natural theology on the idea of laws of nature is far less productive than more promising alternatives – such as those that characterize the pre-modern natural theology of Augustine, Anselm and Aquinas and those new forms now emerging that embrace alternative conceptions of order.

In the second section, I suggest two ways in which the new thinking about order may fruitfully resource a natural theology. Here, I point to two different sets of arguments that draw on ideas of order after the laws of nature. First, for some, the very possibility of natural order is puzzling without reference to God as provider and sustainer of such order. The idea of mindless matter as ordered is so paradoxical that a theistic explanation makes better sense, without having to invoke problematic concepts of the laws of nature or specific divine actions. Second, the venerable argument from design takes on a new lease of life in the context of these new ideas about natural order. Unlike previously dominant models of design that focus on organisms, mechanisms or indeed systems, a new form of design argument is enabled that takes the simple fact of orderliness itself as constitutive of the evidence of divine design.

Natural theology after the laws of nature

There is a widespread consensus that the project of natural theology has definitively run into the buffers and that any attempt to develop arguments to prove the existence and nature of God without reliance on revelation or a particular religious tradition is futile. This consensus view is, however, profoundly mistaken. It is so dominant though that it is worth spending a few moments briefly unpacking the claim – all the better to refute it.

The standard narrative here holds that natural theology has been subjected to a three-pronged attack: philosophical, scientific and theological. As we have already seen in Eric Watkins' chapter (Chapter 1), the key philosophical critiques of the possibility of natural theology are associated with the two great iconoclasts of early-modern philosophy: Hume and Kant. In his *Dialogues on Natural Religion* (1779), David Hume lambasted all attempts to move from claims about the natural order to claims about God (it is, of course, important to remember that Hume's agnosticism extends as equally to those who wish to deny the existence of God on the basis of claims about the natural order as it does to those who wish to affirm His existence). Similarly, in his *Critique of Pure Reason* (1781), Immanuel Kant exposed the folly of speculative claims about knowledge of the existence and nature of God as examples of human reason over-stepping its natural limitations.

Of course, it is undeniable that both these attacks on the very possibility of natural theology are important and must be reckoned with; and yet it is far from obvious that they have really proved fatal to that enterprise. Historically, William Paley's 1802 *Natural Theology* is an extended reply to Hume's 'careless scepticism' and in the mid-nineteenth century natural theologians such as William Whewell (the inventor of the term 'scientist', amongst much else) successfully integrated Kant's insistence on the limits of reason into his natural theological project. Conceptually, it is interesting to see that for both Hume and Kant it is the dependence of natural theology on ideas of the laws of nature that is so problematic. Both challenge what they consider to be the hubris of using ideas of laws of nature as a kind of ladder to ascend from the natural to the divine. Instead of being signs of God's divine authorship of the world, laws of nature are repositioned as human inventions or constructions that tell us only about nature and ourselves and nothing about God. Interestingly though, neither fundamentally challenge the very idea of laws of nature that is put to such – in their view – problematic use by the natural theologians. When that shift occurs, as it now has, then the basis of the Humean-Kantian critiques of natural theology is undermined and a new opportunity opens up for those natural theologians seeking to move from the natural to the divine.

So much for the first prong. The second is the more popularly influential: Darwin. Sitting imperiously enthroned in the temple of the natural sciences (as he does in London's Natural History Museum, for instance), Darwin looms large over the history of the decline of natural theology. Whilst many do accept that it may not have been Darwin's intention to defeat the arguments of natural theology, it is widely held that his theory of evolution by natural selection does precisely that. In light of Darwinian theory it becomes clear that nature feigns design: what appear to be purposely designed natural phenomena are revealed instead as 'simply' the result of a random series of mutations. No longer the guiding hand of God but now the blind watchmaker of evolution is held to be responsible for the state of the natural world. This is a powerful rebuttal of the claims of the natural theologians that continues to have a wide appeal, as the reception of the works of Richard Dawkins clearly shows. Here again, however, the picture is not quite so clear cut.

On the one hand, there have been many, from Asa Gray onwards, who have welcomed Darwin's theory as a salutary incentive to shift the focus of natural theological reflection from particular organisms or mechanisms to the systems of which these individuals are parts. By turning toward a view of nature as a series of intersecting systems, such natural theologians in effect relocate divine design in a way that is perfectly compatible with the Darwinian view of life. Known as 'ramification', after Darwin's own image of the system of nature as a tree of life, this move rejects the destructive either/or of evolution/design and paves the way for a modern Darwinian natural theology, such as, for example, that proposed by Simon Conway Morris.

However, as Martin and Bechtel show in Chapter 4, we should be wary of taking the Darwinian theory of evolution as the sole ordering principle of the biological sciences. As such, perhaps those ramified natural theologies that accommodate Darwin are over-invested in a conception of order in terms of laws of nature that unwittingly entrench the exclusive either/or of evolution or God. Remove the tendency toward viewing evolution as a biological law of nature but rather instead consider it as an ordering strategy of the biological sciences (amongst others) and a new form of scientifically legitimate natural theology emerges.

The final prong on the anti-natural theology fork comes, perhaps surprisingly, from theology itself. Primarily associated with the Swiss reformed theologian Karl Barth, the theological objection to natural theology insists that there is no way from nature to God and that apart from revelation all natural theological speculation is unavoidably idolatrous. There is, of course, an important point here and we must be careful not to let our ideas of God become simply reflections writ large of our own natural concerns. And yet, as Keith Ward emphasizes in his chapter, how we think about God cannot be wholly isolated from how we think of ourselves and the world we find ourselves in. The exalted idea of a 'pure' theology of revelation is thus set in question and the legitimacy of a theological engagement with nature is affirmed. Here, though, care needs to be taken to ensure that such an engagement is a genuinely constructive and mutually enriching one.

A helpful distinction here is between natural theology and so-called 'theology of nature'. The first takes its clues from nature to develop arguments for the existence of God; the second rather 'reads' the natural through the lens of a particular religious tradition. A theology of nature is an account of the natural in light of certain religious commitments. As such, its account of natural order would reflect a prior theological interpretation of the idea of order (perhaps Trinitarian or providential, for example), which then legitimizes scientific and philosophical concepts. This is an interesting theological exercise and there may well be many fruitful convergences between the new understandings of order after the laws of nature and theologically informed reflection on order (just as there were undoubtedly many constructive, as much as there were unfortunate, correlations between theology and the idea of laws of nature). This, however, is not the concern of this chapter.

Interestingly, although the debate is rarely framed in these terms, once again the idea of laws of nature is significant here. Part of Barth's concern about natural theology is that it can, in effect, elevate the 'merely natural' into matters of ultimate significance. As Montuschi and Harré show (Chapter 6), the movement from ideas of natural order to those of social, moral and political order is a short step and it is easy to see how the further step to divine order can be taken. In this, the idea of laws of nature provides a particularly efficient means of over-stepping the boundaries of the natural, social and theological – as indeed the history of the concept of natural law in theology shows. Barth was surely right to worry about the ease with which natural theologians glide to the affirmation of particular natural phenomena as divinely ordained, be that a certain natural order of human relations or a specific political ideology. By qualifying, however, the legitimacy of the idea of laws of nature and re-conceiving ideas of order, this anxiety can, I suggest, be addressed. If ideas of natural order highlight the piecemeal and the local, then the scope for such idolatrous grandstanding seems limited. Thus, here again, the prospects for a new and reinvigorated natural theology seem good.

Two new arguments from order

What might such a natural theology of the new conceptions of natural order after the laws of nature look like? I suggest, in what follows, two ways in which natural theological arguments might proceed.

The most direct is an argument from the very possibility of natural order itself. For this argument there is something inherently puzzling about the very idea of natural order at all. How can it be possible that the motions of mindless unthinking things should behave in an orderly fashion? What is to stop some bits of matter from moving in ways that are disordered, or in the strong sense chaotic? What the new thinking about order after the laws of nature stresses again and again is that order *is* messy and resists our attempts to fix it into the universal, exceptionless and all-encompassing laws so beloved by the early-modern philosophers, scientists and theologians. But order itself remains. It seems here that we are gaining a clearer view of the character of natural order in all its squirming dappledness and yet without the prop of the idea of laws of nature we are confronted with a deeper sense of the paradoxical puzzle of natural order, namely that it in itself seems unfathomable. This is a peculiar and, for some, troubling situation: the very intelligibility of nature that the new thinking about order articulates (and celebrates) seems itself unintelligible, or at least without reason.

Two obvious alternatives assert themselves at this point: either the orderliness of nature is necessary – it just is the way that the world is – or the orderliness of nature is accidental – it just happens to be the way that the world is. Both, however, are unsatisfactory. In the first case, the puzzle of the possibility of natural order simply stays unsolved. How can it be that unthinking, mindless matter behaves in orderly fashions? How can that be possible, let alone necessary? To affirm that it is necessary is to assume that it is possible, which is precisely the question at hand. In the second case, there is obviously something highly problematic about explaining order in terms of randomness without in some way undermining the idea of order that is being explained. To make order accidental is, in effect, to dis-order order.

The natural theologian will see here an argument for the existence of God. What the puzzle about natural order shows is that it is incoherent to suppose that nature will behave in orderly fashions on its own accord. Thus, the possibility of natural order must be explained in another way: by the direct involvement of God. The natural world exhibits order because God provides and sustains it as such. God does not act indirectly through the intermediaries of the laws of nature, setting himself up at a distance like some cosmic watchmaker, winding the mechanism and retiring to self-contemplation. Rather, God is actively present to nature as its orderer.

The second form of natural theological argument suggested by the new thinking about order is a revision of the more traditional argument from design. Whilst the first argument focuses on the possibility of natural order, this second argument looks instead to the apparent fact of order. As is well known, arguments from design have a venerable history, tracing back at least to Plato, and versions of this argument reoccur in almost every period. However, the dominant form has been that developed in the context of the early-modern scientific revolution and the growth of technology. It is not unsurprising that this coincides with the rise of the idea of the laws of nature. According to these 'classic' early-modern design arguments, the apparent analogy between natural phenomena and human artefacts underlies the claim that nature is the product of divine design. As William Paley's famous example has it: just as a pocket watch shows evidence of its human designer, so natural objects show evidence of their divine Creator.

This version of the argument from design prioritizes law-like regularities and instrumental fittedness, consistent with the idea of order in terms of laws of nature. However, it is far from the only form of the argument from design to have been defended in the scientific era. In fact, it is possible to trace something of a development of the argument from design, in which the focus of enquiry shifts over the centuries from the seventeenth to the nineteenth. In broad terms, the movement is from an emphasis on the design of organisms or creatures, to the design of the mechanisms that might be said to constitute such organisms and which are exposed by the works of science. Finally, the argument shifts again to the design of the systems that might be said to sustain the mechanisms. To return to Paley's example of the watch – this is a move from the watch itself first to a focus on the cogs and wheels that make it into a watch, and second to the principles of movement that make the cogs and wheels into a machine that can keep time. This progress of the design argument of course mirrors the progress of the natural sciences toward ever more precise and fundamental statements of the laws of nature. Thus, the fate of the argument from design seems tied to that of the quest for the laws of nature. What then if such a quest is abandoned and a new idea of order put forward?

At first sight it might seem that there is no future for a design argument without laws of nature. As we have seen with the case of Darwin's theory of evolution by natural selection, exponents of the design argument were able to proceed precisely by making Darwin's theory into a form of a law of nature, according to which God enacts His purposeful design. Design, it seems, has become synonymous with ordered regularity and law-like production. Abandon the laws of nature and you abandon the design argument. Indeed, according to the new thinking about order, order is messy and piecemeal: hardly designed order at all.

However, of course, this would be to mistake one form of design for the only possible form. Historically, design has been characterized in ways other than mechanical or systemic regularity and there is no reason why the patterns of order that characterize nature after the laws of nature cannot be considered as evidences of design. For instance, as Bishop and Frigg's chapter shows, order can be identified with relatively stable macroscopic patterns, such as self-organizing systems. These patterns are not explicable solely in terms of laws of nature and thus are not open to being thought of as products of a divine designing intelligence that follows only universal laws. Rather, they could be thought of as resulting from the intimate engagement of the divine in the local particulars. Just as we would find it strange to think of a human designer who simply imposed her blueprints onto whatever material she had at hand without reference to its situation, so too we might favour the idea of God as an involved and situated designer, rather than a dispenser (Henry Ford-like) of one size fits all productions via rigid and unresponsive laws of nature. In other words, the kind of localized messy order might well be precisely the kind of order that would characterize a divinely designed natural order.

Conclusion

Nancy Cartwright

This book has been about science without laws. A long and venerable tradition supposes that science aims to discover laws of nature and to figure out how to deploy them, both to understand the world we live in and to change it. But much of modern science manages to explain, predict, intervene and produce new technologies without recourse to laws of nature. How is this possible: how do the sciences function without employing laws of nature? And how did we come to think laws of nature matter so much in the first place?

The book has had three main aims. First, there may not be laws at work everywhere and every when in the universe at all scales of size and complexity, but there certainly isn't chaos. Much of nature is ordered, systematic, predictable. How can we account for order without laws? To address this question we have looked at recent ways of explaining order without resort to law across different scientific disciplines: in physics, in biology and in social science. Second, the book has aimed to provide the historical and theological setting for thought about laws of nature. How did we come by this concept and how has it shaped and been shaped by developments in the sciences on the one hand and by changing theology doctrines on the other. Third: What are some of the implications of the shifting roles of laws in science for central moral and theological doctrines? For instance, 'Does a universe without law make free will more problematic or less'? Or 'What views of a creator God does the new view of natural laws allow or suggest'?

We are at a new and exciting stage in the scientific and philosophical understanding of nature, and undoubtedly there will be many changes in the near future. We hope that the contributions in this book have shown that the old classical Newtonian view of universal and exceptionless laws of nature is no longer plausible. A view of nature as a realm of diverse powers, potencies and dispositions, creatively building up emergent systems of order, is one that has the potential to reveal new insights about the natural order that will prove fruitful, that will suggest new lines of research and that will resolve some of the old problems about the compatibility of physical causality and the life of the mind that have proved so intractable in the past. For if we are right, we live in a world of order after the laws of nature.

The form that such intimate engagement might take is open to further discussion, but a 'dappled world' view helps to undermine the mistaken idea that divine creation is simply the beginning of a universe which then proceeds on its own according to universal laws. Divine creation has always been viewed theologically as the dependence of every space and time upon some intelligible reality beyond space and time, not just as a divine act of 'lighting the blue touch-paper' to start the universe going, as Stephen Hawking has described it. Thus divine design may be seen not only in the selection of the right initial conditions for a universe which is to generate intelligent life-forms, but also in the direct and continuing dependence of new sets of local conditions upon an ultimate informational and structuring principle of cosmic reality.

Such a view may become more plausible as we reflect on the variety of types and levels of order we find in the world, for instance those represented in the principles of our various sciences, from those we are more accustomed to thinking about in physics, biology and the social sciences to the kinds of order described in this book. The numerous orders that these principles represent may be local and limited and easily disrupted. Nevertheless it requires a vast co-ordination to ensure that they can all live together in one consistent world. One might suppose that God intends the larger scale orders represented in the principles of biology and social science to emerge naturally as the world evolves. In that case huge care is required in the choice of powers and properties, many of them emergent and partly self-organizing, to ensure that the vast variety of levels and kinds of order we see occur.

It is not an aim of this book to reinstate a design argument for the existence of God. But it is part of our concern to ask how a 'dappled world' view of cosmic order may impact upon the beliefs of those who do see God as the intelligent Creator of the universe. In that respect, it may well seem that the amazing – necessarily consistent and integrated – amount and variety of different kinds and different pockets of order, often overlapping, calls out for design even more than a world with one single kind of clockwork order repetitively appearing in accord with one single set of eternal principles.

Index